Reconnaissance

and

ENERGY VISUALIZATION,

STEM Edition

IR Night Recon in Vietnam

Agent Orange to Climate Change

By Stephen E. Weber

Copyright © 2021 by Stephen E. Weber.

All rights reserved. No part of this publication may be reproduced, distributed, or transmitted in any form without the prior written permission of the author or publisher, except in the case of brief quotations in reviews, professional journals, magazines, newspapers, television or radio and certain other noncommercial uses permitted by copyright law. For permission requests, write to the publisher.

ISBN: 978-1-7369215-1-7 (Paperback)
ISBN: 978-1-7369215-2-4 (Hardcover)

Library of Congress Control Number: Applied for March 2021

This book is an informative STEM science book. No liability is assumed for any reference to historical events, facts, people, places or material references whatsoever.

Front cover 1964 complimentary photograph from McDonald Douglass of RF-4C Phantom II Reconnaissance Aircraft.

Book design by Author Stephen E. Weber.

Printed in the United States of America.
First printing edition March, 2021. STEM Edition Sept 2024

Stephen E. Weber, Author / Publisher
Chicago, Illinois, 60656 and Lake Geneva, Wisconsin 53147

Publisher's Cataloging in Publication Data
Weber, Stephen E.
Reconnaissance and ENERGY VISUALIZATION.
IR Night Recon in Vietnam, Agent Orange to Climate Change.

Reconnaissance and ENERGY VISUALIZATION

INTRODUCTION   By Stephen Weber

This is a STEM book: Science, Technology, Engineering, and Math. This book has 4 parts: Reconnaissance, Visualization, Extensions to our vision, and Complex Systems and Problems. Each part is a book in itself and can be studied individually.

Part I: Aerial Reconnaissance. A personal story, the author gets a draft notice, joins the Air Force and is trained in RF-4C aircraft reconnaissance systems and is sent to Vietnam. Chapters 1-11 are Technology and Chapter 4 is Engineering.

How can you make science reading interesting? Perhaps you could make it a personal story. Perhaps you could structure a science story like a novel or a movie. Let us consider a classic film story: "It's a Wonderful Life." It was directed by Frank Capra, released in 1946 and starred Jimmy Stewart. The timeline was the great depression and WWII. The people were poor and could not save enough money to buy a house. There were many farm and home foreclosures. The building and loan allowed a family to purchase a house with monthly mortgage payments.

Consider a science story, a personal experience about Infrared Reconnaissance in Vietnam in the Air Force, Air America and Ranch Hand spraying of Agent Orange leading to the discovery of an energy transformation. High frequency light radiation from the sun is converted to heat energy to warm the earth and longer wavelength lower frequency infrared (IR) radiation emitted by the earth after being warmed by the sun. Global warming and climate change consequences are caused by greenhouse gasses in our atmosphere. The escaping IR energy is absorbed by carbon dioxide, methane, water vapor, nitrous oxide, and ozone and reflected back down towards earth resulting in global warming. The story progresses from recording IR radiation in Vietnam, Agent Orange devastation and defoliation, energy visualization in art, photography, television, telescope, medical imaging and visualizing the results of global warming and climate change 60 years later.

Part II: Visualization. This part is about light energy, Einstein and the photo electric effect and the electromagnetic spectrum. Chapters 12, 13 and 14 are Math. Chapters 18 and 19 are about the Science and psychology of how we see. Chapters 15, 16 and 17 are about flying Technology. Chapters 20 through 24 are about Art. This introduces STEAM: Science, Technology, Engineering, Art and Math. The visual artistic elements of composition are analyzed in music and literature.

Part III: Extensions to Our Vision. With the help of telescopes we can see further away than ever imagined even new galaxies and universes. Microscopes enable us to cellular data even DNA activity. We can view invisible energy from X-rays on film and/or television. There are medical imaging systems such as CT, MRI and PET systems that facilitate viewing internal information in slices for detailed analysis. In addition there are nuclear medicine systems. Chapters 25 thru 34 are about both Science and Engineering

Part IV: Complex Systems and Problems. Global warming and climate change is the big problem presented in part IV. Is it reversible? Reconnaissance is used to measure the rate of change of ice in the polar ice caps and Greenland. Chapter 35 is about Math and calculates the solar radiation. Chapter 36 to 39 is about weather, transportation and global warming and climate change and would be considered Science and Technology. Chapter 40 is new about the James Webb Space Telescope launched by NASA in 2022 and is about Science. It uses infrared technology and IR part of the electromagnetic spectrum. Thus we travel from hi res AF IR recon in Part I to hi resolution NASA James Web Space Telescope in Part IV.

Reconnaissance and ENERGY VISUALIZATION

By Stephen Weber

FORWARD

Air Force tech school in aerial reconnaissance, and worked in night infrared reconnaissance in Vietnam on the RF-4C Phantom II aircraft over 60 years ago. Aerial reconnaissance uses sophisticated equipment to enable us to visualize beyond our capability. The IR sensors take pictures in the infrared region in complete darkness at night. Microscopes can extend our vision to show the cellular structure of plants and animals beyond our original comprehension. Telescopes can enlighten our understanding of the planets and stars in our galaxy and other galaxies. Systems such as these can extend our visual understanding of our universe. X-Rays and CT Scans can help us to visualize and understand what we cannot see and help doctors diagnose problems.

Light from our sun, 93 million miles away, warms the earth and enables us to see. When the earth is warmed by the sun, infrared radiation is given off. According to the Law of Conservation of Energy in order to keep the temperature of the earth constant, the amount of energy from the sun should equal the amount of energy leaving at night. However, all the IR energy cannot escape from our atmosphere because of the green house gasses which leads to global warming and climate change. I wanted to show the knowledge link between IR night recon and global warming due to the IR radiation and greenhouse gasses.

The purpose of writing this experimental book is to try to research, understand and answer these problems. I have tried to make this book interesting and worthwhile by adding Solar Radiation, Weather, and Global Warming. I hope you read and enjoy this journey.

DEDICATION

This book is dedicated to our VVA & VFW veterans who honorably served our country and to my Wisconsin cousins.

The Niles Memorial VFW Post Service Officers (from L-R) Chaplain Steve Weber, Robin Miller, Vince Giuffre, Bob Kosmicki, Tom Fronzak, Frank Merkendorfer, Larry Pike, Bob Fuggiti, Tom Davidson and Commander Harry Benjamin.

Cousins' Lunch, March 21, 2019
Sunset Bar and Grill
Fort Atkinson, Wisconsin

Reconnaissance and ENERGY VISUALIZATION

by Stephen E. Weber

Table of Contents

Part I. Aerial Reconnaissance Systems.
    Chapter 1. A National Crisis and a Draft Notice.
    Chapter 2. Air Force Technical School.
    Chapter 3. Tactical Air Reconnaissance Center.
    Chapter 4. RF-4C Infrared Reconnaissance System.
    Chapter 5. RF-4C IR Recon at Mountain Home.
    Chapter 6. Auxiliary Data Computer System (ADAS).
    Chapter 7. RF-4C IR Night Recon in Vietnam.
    Chapter 8. The Bob Hope Show in Vietnam.
    Chapter 9. Electronic Countermeasures, ELRAC.
    Chapter 10. R & R Japan to Visit NCOIC.
    Chapter 11. Teaching OJT, Agent Orange, TDY.
    Part I. Acknowledgements.

Part II. Light Energy, Aviation and Visualization.
    Chapter 12. Light Energy.
    Chapter 13. Prism and Electromagnetic Spectrum.
    Chapter 14. Einstein- The Photoelectric Effect.
    Chapter 15. Private Pilot Certificate.
    Chapter 16. Aviator Friends.
    Chapter 17. Flight Instructor Certificate.
    Chapter 18. Visual Information Processing.
    Chapter 19. Parallel Processing and Color Vision.
    Chapter 20. Drawing.
    Chapter 21. Engineering Graphics.
    Chapter 22. Painting- Impressionists.
    Chapter 23. The Basic Elements of Design.
    Chapter 24. Artistic elements in music and literature.
    Part II. Acknowledgements.

Reconnaissance and ENERGY VISUALIZATION

by Stephen E. Weber

Table of Contents

Part III. Extensions to our Vision.
Telescopes, Astronomy and Microscopes.
    Chapter25. Galileo and our Galaxy.
    Chapter 26. Yerkes Observatory.
    Chapter 27. Hubble and our Universe.
    Chapter 28. Compound Microscope.
    Chapter 29. Electron Microscope.
Photography, Television and Medical Imaging Systems.
    Chapter 30. The Camera and Photography.
    Chapter 31. Motion Pictures.
    Chapter 32. Television.
    Chapter 33. Medical X-Ray.
    Chapter 34. Linear Accelerator, CT and PET Scan.
    Part III. Acknowledgements.

Part IV. Complex Systems & Problems.
    Chapter 35. Solar Radiation.
    Chapter 36. The Weather.
    Chapter 37. Energy and Transportation.
    Chapter 38. Global Warming and Climate Change.
    Chapter 39. Entropy and Thermodynamics.
    Chapter 40. James Webb Space Telescope.
    Part IV. Acknowledgements.

Appendixes.
    Appendix A. "You Are Not Alone." Speech
    Appendix B. Computer Systems Consulting Service-
        Educational Proposals.
    Appendix C. Agent Orange Speech.

Bibliography.

Recnnaissance and ENERGY VISUALIZATION

By Stephen Weber

Part I. Aerial Reconnaissance Systems.

Chapter 1. National Crisis, Draft Notice and Basic Training.

President Kennedy was shot in Dallas, Texas on November 22, 1963. Several days later my draft notice arrived in the mail. The Lincoln Continental Convertible carrying the President, the First Lady, and Governor Connolly of Texas passed by the Texas Book Depository on its way to where President Kennedy was to give a speech. Three shots rang out and President Kennedy slumped down in the limousine as it sped away to Parkland Memorial Hospital emergency room where Dr. Malcolm Perry was on duty. According to the Saturday Evening Post an endotracheal tube was placed down his throat to start oxygen. Dr Perry started a tracheotomy and a chest tube to suction out blood. A transfusion was started with type O-negative blood. Dr. Perry massaged the chest for ten minutes to stimulate the heart. Since there was no breathing and no heart beat, President Kennedy was pronounced dead at 1pm. Father Huber anointed John Kennedy with oil, absolved him of his sins and Catholic Last Rights were administered. All this time Jacqueline Kennedy was with her husband in the emergency room. Everything happened so fast. Vice President Lyndon Johnson was sworn in as the next president of the United States.

The news was on the radio and television around the country and around the world. The police rushed into the building to the 6th floor window where the shots were fired, just as Lee Harvey Oswald was leaving the Texas Book Depository. Oswald went to his apartment to get a jacket and ran out. A patrolman, J.D. Tippitt, recognized him from the description and tried to stop him. Witnesses said Oswald shot the patrolman and went into a movie theater. Some witnesses ran into the theater after him. Oswald was arrested and brought to a Dallas police station, where he denied any guilt and pleaded innocent. The next day in a police corridor Jack Ruby, a strip club owner, shot Lee Harvey Oswald. He died at Parkland Memorial Hospital 2 days later and 10 feet from where Kennedy died.

There were many unanswered questions. Some bystanders said they heard several shots fired from behind a nearby grassy knoll. Rifle experts said that it was impossible for Oswald to get off three shots with accuracy that fast with that old Italian bolt action rifle. Did Ruby shoot Oswald to keep him from talking? Was this a communist plot? Was the CIA involved? My cousin Tom Gayne was a student at Gordon Tech High School in Chicago. They had closed circuit television and watched the whole incident repeatedly unfolding on television every day. They loved President Kennedy. After graduating from college, my cousin Tom went back to teach at Gordon Tech. My cousin Tom researched and became a Kennedy scholar and believes that Oswald was in the December 14, 1963 Saturday Evening Post picture and that he was set up as a fall guy. Tom also believed that this was a CIA plot. Kennedy reduced US troop buildup in Vietnam.

According to <u>Vietnam: A History</u> by Stanley Karnow: In 1961 Kennedy set up an economic, social, political and military task force to prevent communist domination of South Vietnam. Kennedy followed Eisenhower's strategy and sent in military advisors. Kennedy opposed the introduction of combat troops in Vietnam. President Ngo Dinh Diem and his brother Nhu were murdered on November 2, 1963 in the Diem coup. Several weeks later President Kennedy was killed on November 22, 1963. Most people in the US loved President Kennedy. This was a national tragedy. Is it any wonder why thousands of draft notices were sent out right afterwards? Was

I shocked? What a tragedy! What should I do with my draft notice? I really had to act fast.

When I received my Army draft notice, I went to the Air Force recruiter and said "My cousins were in the Air Force. Can I get in the Air Force?" They said "Ya gotta take a test." I showed them my draft notice and said, "I graduated from Wright College". They said "You still have to take a test." Fortunately, I passed and went to basic training boot camp and tech school in aerial reconnaissance shortly after the assassination of President Kennedy. After being sworn in, the other recruits and I took off from Chicago's O'Hare Field and flew to San Antonio and on to Lackland Air Force Base to process in, get chewed out by our first sergeant and have some chow. The next morning we were given green fatigues and brogans and started marching and drilling. We had indoctrination films and tests during the day and more marching and drilling in boot camp basic training. In the afternoon we went to the rifle range and did more marching. Although we went to bed early and got up early, we were exhausted.

We did not get that much sleep, because every couple hours we were awakened by a fire drill or a cry of "gas, gas, gas" and have to form in the athletic field to move telephone poles from one end to the other. We had about 50 men to a flight with 4 or 5 squads. Each squad would have to carry a telephone pole to the other end of the football field. We did this almost every night. Everyone was tired. The concept of basic training was to break you down and then build you up to work as a team. Several times a week we were on the rifle range. Finally we got measured for our Air Force blues, summer tans, overcoat, dress shoes and fatigue jacket. After 5 weeks we were allowed to go to town in San Antonio as gentlemen and visit the Alamo.

After graduation, we got our orders for tech school. I was given Lowry Air Force Base in Denver, Colorado to study electronics and aerial recon systems. We had a long two day train ride on the Katy line to get there. It took over 30 hours just to get out of Texas. We went from San Antonio to Dallas then west to El Paso and north following Route 25 through New Mexico and then to Colorado Springs and finally to Denver, the mile high city. Because of the altitude , we were told to limit ourselves to one beer. In January it was ice cold in Denver. We were freezing. Somehow we got to Lowry Air Force Base with our heavy duffle bags and found our new barracks. Lowry is located between Aurora and Denver and just south of Stapleton International Airport. Lowry AFB was an Air Training Command base 56 years ago. It was closed along with other bases in the late 80's or 90's and Stapleton Airport is now called Denver International Airport.

We would have an intensive 8 months of training at tech school. We would be the first class to have televised electronics training from 6am in the morning to 12 noon and we would be the last class to cover both ground photography as well as aerial photographic systems. They said we had to study hard because there were a lot of new systems coming out besides radar and television that we did not know anything about. It sounded exciting. What didn't we know?

Figure 1.1 Vietnam, A History by Stanley Karnow

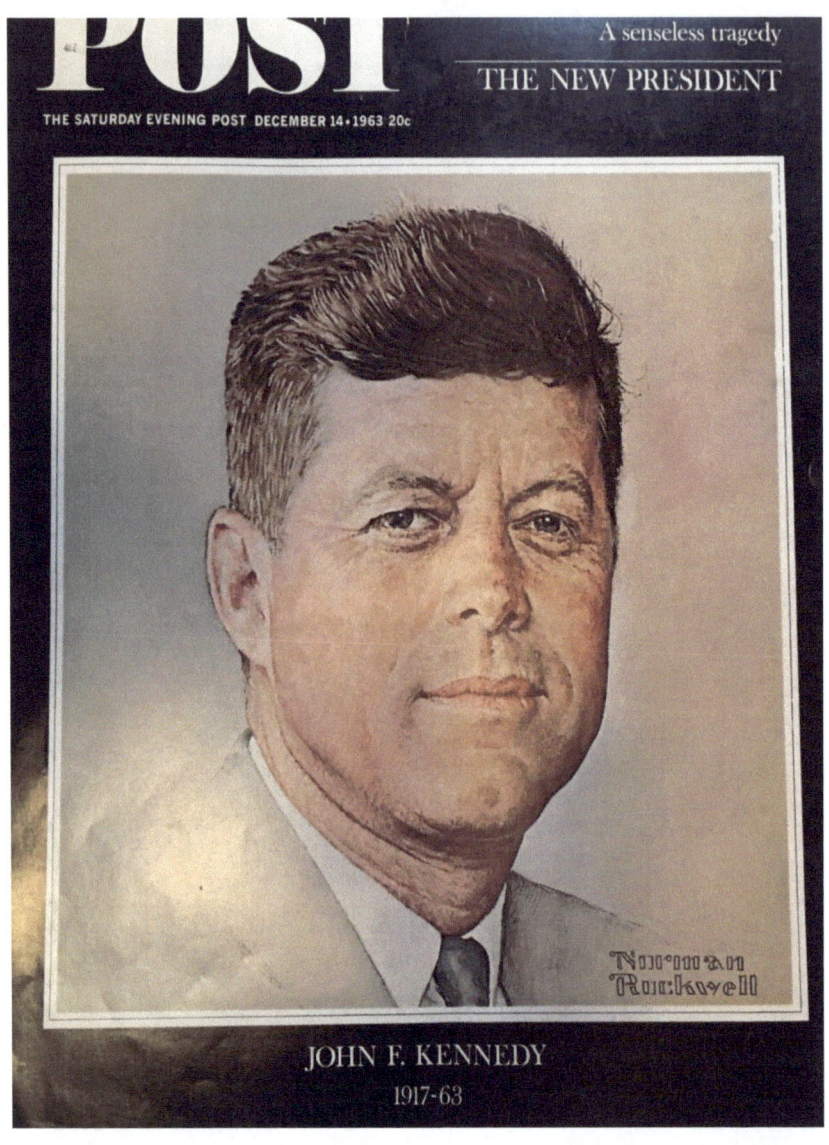

Figure 1.2 Saturday Evening Post, December 14, 1963. Cover

Figure 1.3 Saturday Evening Post, December 14, 1963, Pg.24 The famous Saturday Evening Post photograph snapped a second after Kennedy was shot.

The Lincoln Continental Convertible carrying the President, the First Lady, and Governor Connolly of Texas passed by the Texas Book Depository on his way to give a speech. Lee Harvey Oswald is pictured watching the parade at the top right of the Saturday Evening Post photo. Perhaps Oswald went down to watch the parade and was set up as the fall guy. The resident is leaning forward as his head is partially obscured by the driver's side rear view mirror. They said three shots

Figure 1.3 Saturday Evening Post, December 14, 1963, Pg.25

rang out. The limousine driver immediately pulled ahead of the procession and sped towards the emergency room of the Dallas Parkland Memorial Hospital. This is not the first time a president was shot. President Lincoln was killed almost a century earlier in 1865.

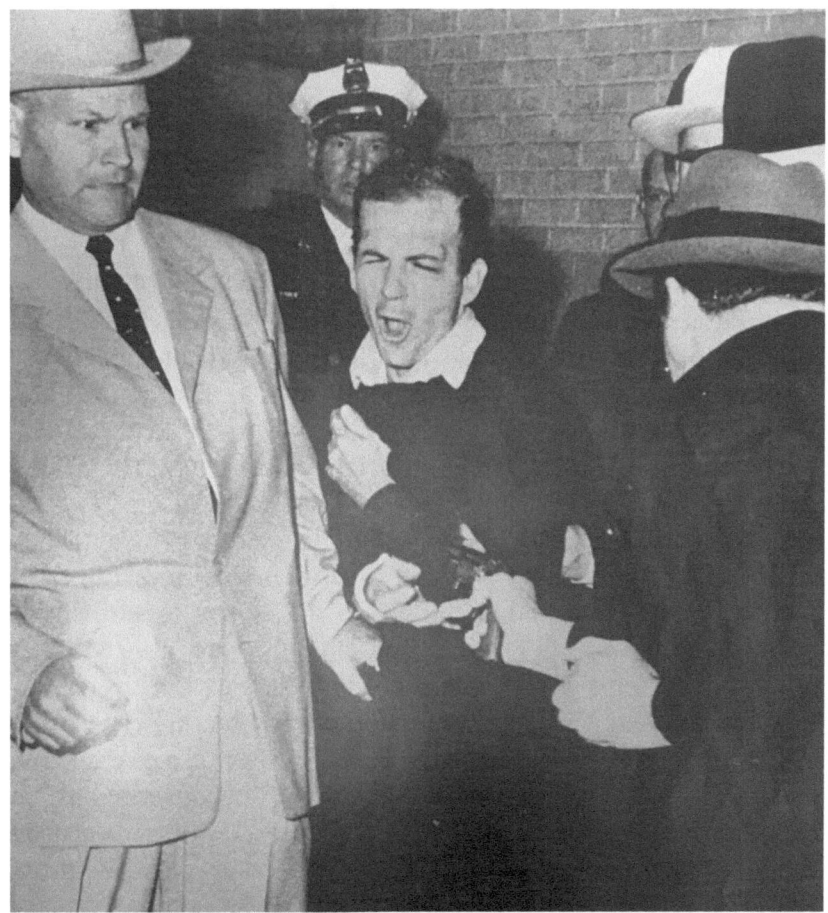

Figure 1.4, Saturday Evening Post, December 14, 1963, pg.29

After President Kennedy was shot, Lee Harvey Oswald was captured and brought to the Dallas Police station. On the next day Oswald was shot by Jack Ruby in the police station. This stopped the interrogation, the investigation, and any further government involvement.

Reconnaissance and ENERGY VISUALIZATION

By Stephen Weber

Part I. Aerial Reconnaissance Systems.

Chapter 2 Air Force Technical School.

Aerial Photo Recon was an 8 month long tech school with 5 to 6 months of electronics and 2 to 3 months of aerial photographic systems. We had electronics technical school from 6am in the morning to 12 noon Monday through Friday. In the afternoons we had electronic labs. Some of the sections were televised. I met James Albright and Charles Gonzalez from Los Angeles.
We had great instructors. We studied both AC and DC electronics, motors and servos, vacuum tubes and transistor theory. We studied radio, television and radar fundamentals because some of the newer recon equipment consisted of televised systems and radar recording cameras. Our classes moved at an exciting pace. We studied and helped each other out so that no one was left behind. We were a great team. We used slide rules for the math calculations because electronic calculators were not available in the early 1960's. We used the latest test equipment in our labs including VTVM's, dual trace oscilloscopes, VOM's, and signal generators.

The systems portions of our aerospace photographic repairman course considered the aircraft systems configurations with cockpit controls, aircraft wiring and aircraft power, gyro stabilized mounts, servo systems, radar altimeter inputs, velocity and airspeed inputs, inertial navigation inputs for longitude and latitude. Simulated problems and solutions were presented and diagnosed. Since the advent of the airplane, aerial reconnaissance has played a decisive advantage in every war including the First World War, where visual information from the pilot provided key strategic information about the enemies' position, supplies, strengths and weaknesses as well as our allies positions for proper support. In addition, gun cameras provided fighter aircraft pilots evidence that an enemy aircraft was shot down in battle. In World War II aerial reconnaissance cameras were used by both sides in bombing missions to assess the damage and in planning future bombing missions for thousands of B-17's, B-24's and B-29's.

After 8 months of electronics and systems schooling we finally graduated in August 1964 and received our diplomas and Permanent Change of Station, PCS assignments. That very afternoon as we were driving home from Lowry AFB in Denver, the Beatles were arriving at Stapleton Field. We listened to the dramatic excitement on the radio. There were two sets of Beatles. The real Beatles were going to stay at the Brown Palace hotel. We loved tech school.

We talked about the excitement of the grand opening of "The Unsinkable Molly Brown" in Denver with Debbie Reynolds and Havre Presnel. Some of the guys remembered the Saturday visit to the Coors brewery while others remembered the train ride to ski at Breckenridge or a ride on the Estes Park tramway. We loved Denver. We will miss Denver, the mile high city, but we are headed to our Permanent Change of Station, PCS Base. My assignment would be the Tactical Air Reconnaissance Center at Shaw AFB to train reconnaissance pilots on the new RF-4C Phantom II Aircraft from McDonnell Douglas. The following are examples of reconnaissance aircraft: RF-4C Phantom II, RF-101 Voodoo, RB-66, U-2, SR-71 Blackbird, and the A5 Vigilante.

Reconnaissance systems really came of age in the Vietnam War because of the new state of the art solid state technologies like the RF-101 Voodoo, the RB-66, and especially the RF-4C Phantom II which was the first solid state all weather day/night Reconnaissance Aircraft. Then there are other specialized intelligence gathering aircraft like the U2 and the YF-12A, SR-71 that gather ELINT information such as radar profiles and uplink data from surface to air missile (SAM) sites. Satellite based reconnaissance systems have increased in capability after the Apollo Moon launch in 1969.

In the late 1980's drone capabilities have increased capability. The drones have CCD TV cameras that can transmit back to the remotely piloted controller who could launch a strike with a missile and determine if the target was hit. Now smaller helicopter style drones with a CCD TV cameras' that can transmit the images back to one's cell phone are being sold commercially to anyone that wants an aerial view of whatever they are researching. Typically these drones have 4 muffin fans, one on each corner with the TV camera in the center. There are flight restrictions especially near airports and licensing requirements for drones. Pictured: RF-4C Phantom.

Figure 2.1 RF-4C Phantom II, Wright Patterson AFB Museum photo with author Steve Weber

Figure 2.2 RF-4C Phantom II, flight line patch decal.

The RF-4C MacDonald Douglas Phantom II is a solid state all weather day/ night reconnaissance aircraft that was a workhorse during the Vietnam War. It was used by the Air Force, Navy and Marines. Pictured author Stephen Weber with the RF-4C aircraft at Wright Patterson Air Force Museum. On the far left is the peto static system which is used to measure the aircraft speed. Aft of that is the forward looking sector scan radar with a radar recording camera. Aft of that is the forward oblique KS-72 camera which uses 5 inch film and takes pictures at the max rate of six frames per second depending on the altitude and air speed the aircraft is flying at. The ACPC or Air Craft Parameter Control receives the air speed from the peto tube and altitude from the radar altimeter and provides a Velocity over Altitude Signal v/h to all the recon sensors on the aircraft. Aft of the forward oblique is the KA-56 low altitude panoramic camera which takes panoramic pictures from horizon to horizon 180 degrees at six frames per second. It uses a rotating prism to paint the picture on the film. Aft of that is the high altitude pan or split verticles. Aft of that is the Infrared Reconnaissance System which takes pictures at night. Aft of that is the side looking radar. The image from the side looking radar is processed by computer.

Figure 2.3 RF-101 Reconnaissance Aircraft, AF photo.

The RF-101 Voodoo is a fast reconnaissance aircraft that was used in Vietnam for day time reconnaissance missions. The RF-101 with it's high airframe could take off and land during the monsoons with the flooded metal clad Air America runways. The RF-101 pilots used aero drag with the recon nose high to slow the aircraft down for landings. It looked like a speed boat taking off and landing during the monsoon rains..

Figure 2.4, Lockheed SR-71, Wright Patterson AF Museum with author Steve Weber.

The Lockheed SR-71 Blackbird is an extremely high altitude reconnaissance aircraft that can gather ELINT information for electronic countermeasures information and provide ECM jamming capability against enemy SAM sites. The Lockheed SR-71 Blackbird aircraft could fly higher and faster than enemy surface to air missiles over North Vietnam and gather important ELINT threat information to make it safer for our reconnaissance flights. The aircraft has a sophisticated electronic counter measures ECM system to effectively jam the enemy SAM sites.

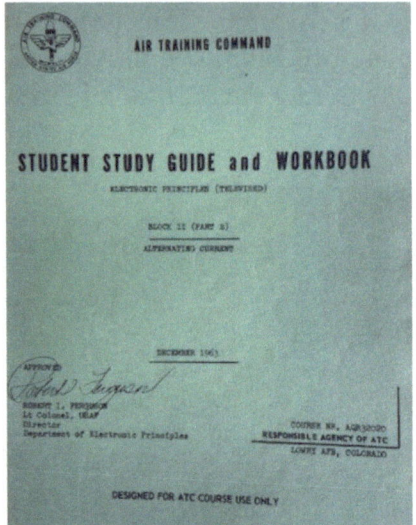
Figure 2.5 Televised Electronics.

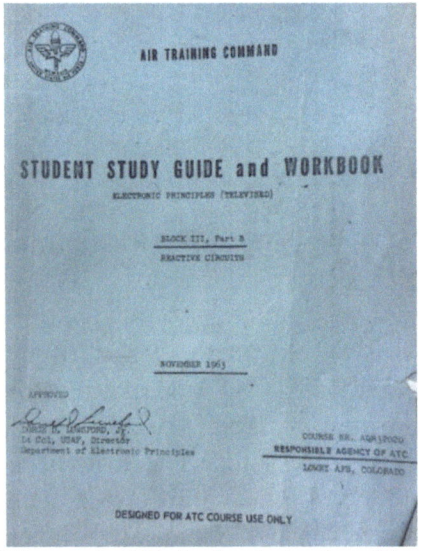
Figure 2.6 Televised Active Ckts.

Reconnaissance and ENERGY VISUALIZATION
By Stephen Weber
Part I. Aerial Reconnaissance Systems.

Chapter 3. Tactical Air Reconnaissance Center, Shaw AFB.

We had several weeks leave after tech school before arriving at the 4411th TARC at Shaw AFB only to discover that the RF-4C aircraft had not arrived from the factory yet. So we started OJT, On the Job Training, working on the RF-101 and RB-66 aircraft recon systems while attending the Field Training Detachment courses on the RF-4C. Chief Master Sergeant Potter was a WWII and Korean War Veteran and the NCOIC of the 4411th Tactical Air Reconnaissance Center. He wanted new airmen to volunteer for Vietnam. TARC's mission was to test new systems and train reconnaissance pilots for Vietnam. He and Sgt Hoover were on flight status. Everything had to be perfect. Sgt Dogherty and Warren supervised OJT. The older experienced airmen: Danny Whalen, Wayne Barren, James Hunt, Dave Devoe, Ed LaRegina and David Williams shared their experience in working on aircraft systems, pre flights, and post flights.

Figure 3.1 RB-66 Reconnaissance Bomber AF photo

Figure 3.2 TAC Aero Club Piper Cub

CMSGT Potter was tough and strict, perhaps because he tried to keep the qualified married airmen at Shaw and deploy the single airmen to Vietnam. Excellence was required. He did this by making promotions difficult and life uncomfortable. There were TDY's (Temporary Duty Assignments). My roommate Dave Asman went to Northrop in California on the Skoshi Tiger Recon Project. He was sent to Ton Son Nut and later came back as a Tech Rep and was killed in Vietnam.

Both Charles Gonzales and I wished we had more time for the TAC Aero Club. They had an aviation ground, flight school and a Piper Cub and a Cessna 150 for instruction. It was rumored they had a very popular 4 passenger Cessna 172. Gonzales, Bernier, Asman and LaRegina worked in the Field Shop, while Albright, Whitaker, Jerry Verrett and I worked the flight line. CMSGT Potter was too strict. There was no time off. Of course we did not have the money for the Aero Club either. Just maybe, that was the plan to get people to volunteer for Vietnam? Gonzalez volunteered for Vietnam and Thailand. Was that a surprise! We were told never to volunteer for anything. Later I would volunteer for IR School.

Figure 3.3 TAC Aero Club Cessna 150.

In the fall of 1964 three RF-4C aircraft, 747, 748, and 749 arrived from McDonnell Douglas. These aircraft had a radar recording camera in the nose. The forward oblique camera was a Hycon KS-72 that could take 6 pictures per second with 5 inch film. From front to rear, the next station was the KA-56 wide angle pan camera that took pictures from horizon to horizon at the rate of 6 frames per second. The pictures were 18 inches long and 5 inches wide. This was accomplished with a rotating prism that painted the image on the film. This system was made by Fairchild and the photograph looked like a football with a vanishing point at either end. The system just aft of that is the high altitude pan or two split vertical cameras with long lens cones.

The next station is the ACPC or Air Craft Parameter Control and the Auxiliary Data Annotation System or ADAS. Information from the Inertial Navigation System INS-12 Recon Adapter is sent to the ADAS which encodes that data digitally and sends it to one inch CRT's in each sensor to be displayed on each picture of each sensor. The specific recon configuration for each sortie is setup on the ACPC. The back seat radar, photo systems and ECM system operator on the RF-4 selects which sensors he wants to operate and that information goes through the ACPC to that sensor. Just aft of this are the next two sensors that operate day or night, the Goodyear Side Looking Radar and the Texas Instruments AAS-18 Infra Red Line Scan Mapping System.

I was asked by Ed La Regina to help coach a little league baseball team that incidentally won the base championship. I volunteered to work night shift so that I could go to the FTD, Field Training Detachment AAS/18 Infrared Reconnaissance System IR school in the day for 5 weeks 8 hours a day. This was a very complex system that would be used extensively in Vietnam at night for IR night reconnaissance missions.

Figure 3.4 RF-101 Voodoo Reconnaissance Aircraft

Figure 3.5 KS-72 Forward Oblique 5 inch film Samples Courtesy Charles Gonzalez.

Figure 3.6 Field Shop Work, Gonzalez and LaRegina.

On a typical high speed 1966 recon sortie over Haiphong the KS-72 forward oblique camera can take pictures on 5 inch film at the maximum $\frac{v}{h}$ velocity over altitude rate of 6 frames per second. This is because of the tractor feed five inch film. Notice the one inch encoded ADAS time, longitude, latitude, and date stamp on the top right.

Figure 3.7 McDonald Douglas RF-4C Phantom II. Picture courtesy and compliments of McDonald Douglas.

This was the first All Weather Day Night Reconnaissance Aircraft. I was all solid state – no vacuum tubes. That was important because of the G forces this aircraft could pull and the vibrations. It had bulkhead wiring connectors and no long cables from front to rear. All systems were fault tolerant. There was mechanical linkage, electrical controls, hydraulic controls and pneudraulic controls. The Infrared System could take high resolution TV line scan pictures at night. The side looking radar system could take pictures through clouds. There were even two cockpits and either the pilot or co-pilot (also called the PSO or ECM operator or radar navigator) could fly the aircraft. Both cockpits had armed ejection seats.

Reconnaissance and ENERGY VISUALIZATION

By Stephen Weber

Part I. Aerial Reconnaissance Systems.

Chapter 4. RF-4C Infrared Reconnaissance System.

The TI AAS-18 IR System consists of 3 main parts with lots of inputs from the aircraft. These are the receiver, the recorder and the magazine. In addition the PSO IR cockpit control unit is used to operate the IR system remotely from the cockpit. The Inertial Navigation System made by Litton Industries and the recon adapter unit provide longitude, latitude, roll, drift and pitch to the IR system and the IR gyro stabilized mount.

The IR system provides viewing 60 degrees on either side of nadir or about a 120 degree view. Objects on the ground are warmed by the sun. These objects function as black body radiators and give off thermal IR radiation. The IR receiver has a scanner rotating at 12,000 rpm's and video from the ground reflects of the front surface optics of the scanner and is reflected up via 2 sets of folding optics to a parabolic reflector. The video is then reflected to a mercury doped germanium detector housed within a cryogenerator cooled to 4 degrees Kelvin above absolute zero. The scanner has 4 sides which yields 48,000 line scans per minute. Video from the cryo cooled detector is amplified by a video preamplifier and then sent to an AGC, automatic gain control amplifier. The video then goes to the recorder.

The recorder functions like a television. Although 48,000 line scans of video per minute sounds like high resolution, a conventional TV has 525 lines of video per frame. There are 30 frames per second and 60 frames per minute. The receiver scanner is 4 sided front surface optics and each ¼ turn creates one line of video which is sent to the recorder. The cathode ray tube in the recorder has horizontal and vertical deflection plates. The recorder electronics generate both the horizontal and vertical saw tooth sweep circuits and the staircase generator waveform. The video signal is applied to the grid and the horizontal and vertical sweep voltages are applied to the horizontal and vertical deflection plates of the CRT. The staircase waveform generator ramps up the deflection plate voltage so that a rectangular raster is formed a line at a time.

The IR magazine contains a 500 foot roll of film that moves at the $\frac{V}{H}$ rate, the velocity over altitude rate coming from the ACPC. The video from the recorder exposes the film quickly as it moves by the recorder CRT. The image is continuous and is composed of fine horizontal lines 5 inches wide. The slower the pilot is flying the closer together the lines would be spaced. For example if the pilot was flying along a river and the IR system was turned on for 5 miles. The film after it was developed would show 5 miles of river on the length of the film with 120 degrees on the 5 inch width of the film depending on the altitude and groundspeed of the aircraft. The IR system takes high resolution thermal TV images, photos, in complete darkness.

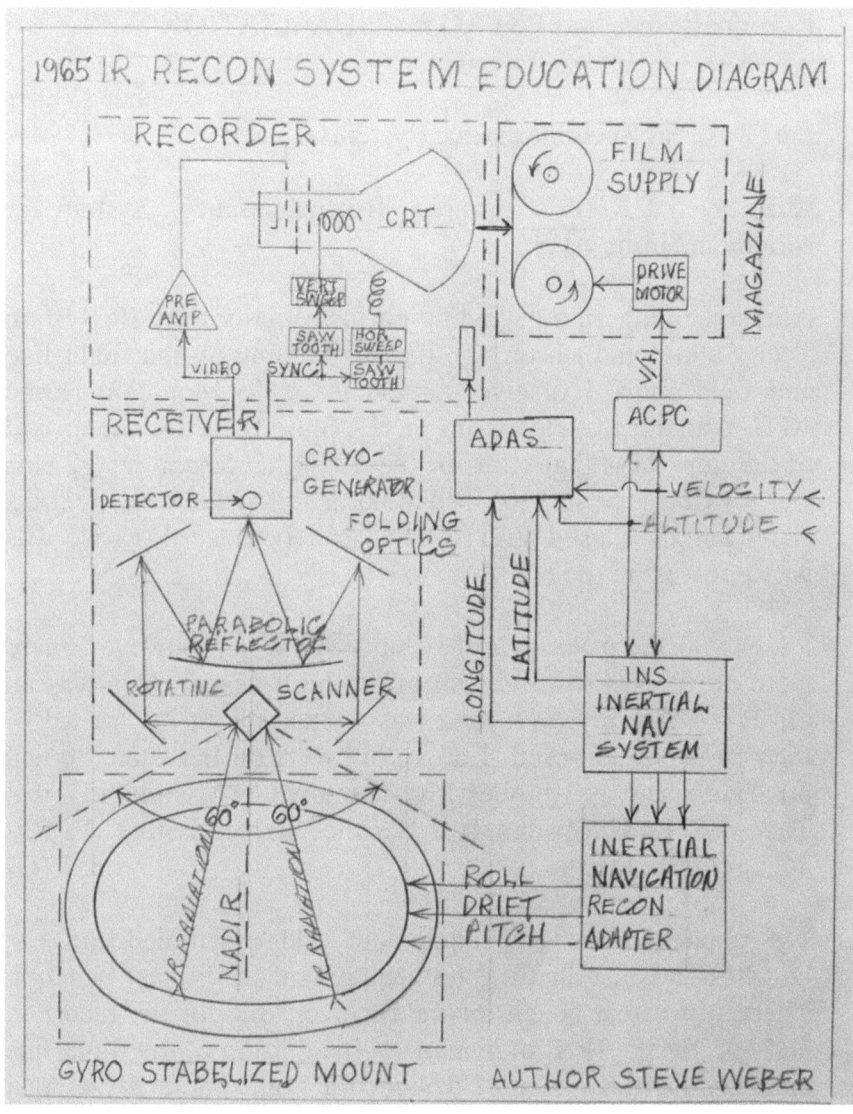

Figure 4.1 IR Reconnaissance System Educational Functional Block Diagram.

Reconnaissance and ENERGY VISUALIZATION

Stephen Weber

Part I. Aerial Reconnaissance Systems.

Chapter 5. RF-4C Infrared Reconnaissance System at Mountain Home AFB.

After receiving an outstanding performance for the IR System Tech School and starting on the job training on the flight line and field shop, I received orders to go to Mountain Home AFB in Idaho to help set up the IR Shop there for the arrival of the new RF-4C aircraft for the 67th TAC Recon wing. This was a PCS move so I stopped at home to say hello to my folks and I bought a used Volvo P1800 car, because this base was way out west in the country.

I was told there was a girl behind every tree however, there were no trees just sagebrush in the high desert. This was an old B-47 base and everyone was excited about the arrival of the new RF-4C aircraft. I drove out west on Interstate 80 and 30. The base was 10 miles from the town of Mountain Home. The AAS-18 IR System Test Benches arrived before the RF-4C aircraft.

We unpacked the test equipment and assembled the test benches and installed the test equipment in the test benches. There was a test bench for the receiver, one for the recorder, and one for the IR film magazine. In addition to the flight line analyzer for the system, I designed a flight line analyzer that would include the helium tank and the pressure gauges. I did the engineering drawings and submitted them for approval. For some reason base engineering said they could build it if I gave them credit for it.

The RF-4C aircraft started arriving from McDonnell Douglas. My future brother in law, Tom Egan, arrived just in time to help with the arrival of the new aircraft. He was well trained in Instruments Tech School at Chanute AFB in Rantoul, Illinois. When I was home on leave after Lowry AFB Tech School, my sister Cheryl introduced me to her fiancé Tom Egan, who graduated from Gordon Tech, a Catholic High School and then DePaul University. I talked him into joining the Air Force if he received his draft notice. We had to inventory, preflight and thoroughly check the aircraft out. I was asked to go to another FTD tech school at Mountain Home AFB, transistor circuit theory.

Figure 5.1 1963 Volvo P-1800 sports car with Steve Weber.

Steve Weber sits inside his 1963 Volvo P1800 spots car at Mountain Home AFB showing the custom console that he designed with black leather, birch plywood, a speaker grill and a hole cutout for the floor mounted 4 speed transmission with a Laylock overdrive for the highway. The console sits on a new red carpet with used Porsche jump seats in the rear of the Volvo coupe.

Reconnaissance and ENERGY VISUALIZATION

By Stephen Weber

Part I. Aerial Reconnaissance Systems.

Chapter 6. RF-4C Auxiliary Data Annotation System (ADAS).

The Tech School was called the "The Aircraft Inertial and Radar Navigation Systems Technician (RF-4C) (Data Display Set AN/ASQ-90). This was an on base class, so we went to school during the day and worked at night. The longitude and latitude data from the INS-12 inertial navigation system along with the radar altitude, the time and the date were sent to ADAS where it was encoded and sent to each reconnaissance sensor to be displayed via a 1 inch CRT on the film.

The problem was that this information was encoded so no one could understand it without a card to decode the information, which took valuable time and a decode card. This would prove to be a great hindrance in interpreting the data on the film for photo intelligence in Vietnam. Several years later when I was working at Zenith's Military Engineering "Video Data Group" a proposal arrived to bid on a proposal to convert the digital output from ADAS to a readable format to be displayed on each sensor. There was no logical reason why it should have been encoded in the first place. The correct information would have really helped senior military officers make quick intelligent decisions, rather than trying to decode the ADAS data slowly.

This was apparent during the Tet Offensive in 1968 and 1969 when I was working at Zenith. The hourly IR missions at night in 1966 and 1967 targeted the Ho Chi Minh Trail as well as the SAM sites in Hanoi and Hi Phong harbor. However, perhaps they did not have the accurate longitude and latitude settings decoded from ADAS. Otherwise they could have stopped the infiltration of the North Vietnamese coming in along the Ho Chi Minh trail. Either photo intelligence did not notice it or the recon pilots did not photograph it at night, every night the movement of people along the Ho Chi Minh trail. Perhaps the missions did not request these flights or perhaps they were not aware of it until the Tet Offensive.

Mountain Home AFB had a great Automotive Center where you could work on your car. I designed and built a mahogany and leather center console for my Volvo P1800 and installed it along with Porsche jump seats for the rear from a junk yard in Boise and a new carpet. Then I gave it a tune up. It was all set for a weekend ride to Sun Valley, Idaho several hours away. Then, I attended another class on solid state electronics, transistor circuit theory for a week. Some of us that came from TARC at Shaw AFB got orders to go to Vietnam and set up in advance of the RF-4C aircraft arrival. We were the RF-4C advance set up and training. In TAC we had a duffel bag packed. I gave Rivera a ride to Chicago and he went to O'Hare Field for a connecting flight to New York for leave before going to Vietnam.

Would you believe they tore up the road between the town of Mountain Home, Idaho and the Air Force Base 10 miles away. It was a two lane road and they wanted a 4 lane highway. Instead of building a new two lane divided highway next to the old 2 lane road, they tore up the existing 2 lane road and went on strike. According to my future brother in law, Tom Egan, there were tires, tail pipes and mufflers scattered around. My sister Cheryl and Tom got married while I was overseas. Cheryl moved to join her husband Tom at Mountain Home AFB and soon my niece Elizabeth was born. Tom was very knowledgeable in aircraft instruments and navigation and went TDY to Nellis AFB in Las Vegas to work on an SR-71 Blackbird and several other aircraft there. This was fortunate that he did not have to go overseas to Vietnam during the war. He even got promoted to Staff Sergeant after Liz was born.

Figure 6.0 a. Aerial Photograph with one inch ADAS time and date stamp. Courtesy Charles Gonzalez

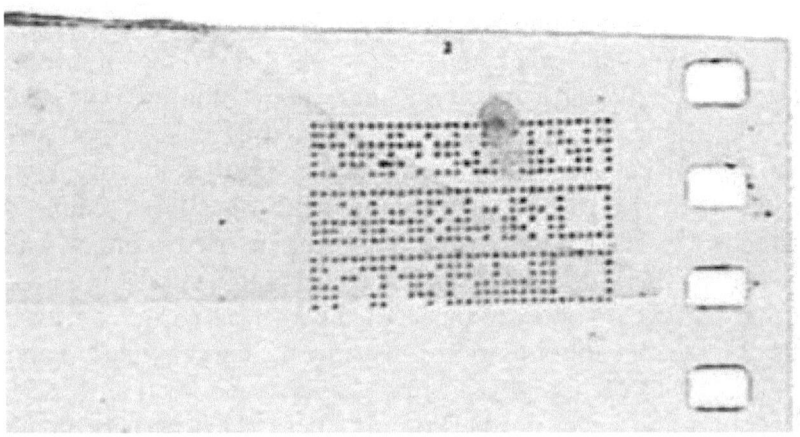

Figure 6.0 b. ADAS Detail with time, date, longitude, latitude, and radar altitude coded.

Figure 6.1 Author Steve Weber with his 1963 Volvo P1800 on July 4, 1966 weekend in Cheyenne on the way home from Mountain Home AFB, after receiving orders for to go to Vietnam. Steve and Rivera drove the 1800 miles to Chicago on Interstate 30 and 80 in 30 hours. Then Steve drove Rivera to O'Hare Airport before going to Vietnam.

Figure, 6.2 1963 Volvo P1800 in front of author's home.

Reconnaissance and ENERGY VISUALIZATION.

By Stephen Weber

Part I. Aerial Reconnaissance Systems.

Chapter 7. RF-4C IR Night Recon in Vietnam.

The time is 1966 and troop deployment to Vietnam were increasing to over 200,000. In several years there would be over 500,000 troops deployed. I went home on leave not knowing what to expect in the future and sold my car. The airlines were on strike and we had orders to go overseas to APO San Francisco. We contacted the Air Force Reserve at O'Hare Field and got a ride on a C-124 from O'Hare Field to Travis AFB in northern California. Then we flew commercial to Hawaii, a perfect climate, to Clark AB in the Philippines, high humidity, to Guam and then Ton Son Neut in Saigon.

Instead of a smooth glide slope, we dropped down for the landing and bounced 20 feet in the air, which is better than getting shot at. We later found out that some of the guys from the 4411th at Shaw along with some airmen from the 363rd went to Ton Son Neut, while some of the 4411th guys along with some of us from the 67th went to a remote Air America base, the closest air base to Hanoi, near Vientain, Laos, Udorn RTAFB in northern Thailand. We got there during the monsoon rains. Everything was under water, even the metal clad runway. We arrived before our RF-4C aircraft. So we worked on the RF-101 Voodoo's while waiting for the RF-4C's to arrive. Then the AAS-18 IR System test benches arrived and we set them up right away on day shift because we were still waiting for our RF-4C aircraft.

We were in monsoon season. There were 4 or 5 major downpours per day. We lived in a Quonset hut on posts with flaps that opened up and screens to let the air in for circulation. Everything was on the water. The walkways were like boat piers. We had house girls that worked during the day only and went home at night. The flight line and the runway were under water. The aircraft looked like speed boats taking off and landing. The RF-4C aircraft started arriving from the 67th at Mountain Home AFB and the 4411th TARC at Shaw AFB. Sgt. George Gallo was the chief of the IR shop, he was an electronics instructor from Lowry AFB ATC and very muscular. He was a good supervisor. Sgt. David K. Williams was night shift supervisor. I was working swing shift some days and some nights depending on the requirements. Then the new requirements came down. The pilots would fly IR sorties every hour all night long as well as day shift. We did not have the manpower to cover this. It would mean working much longer hours. I went from working evenings from 3pm to midnight to working all night from 5pm in the evening to 7am in the morning, 14 hours per day, 7 days per week.

Figure 7.1 Author Steve Weber working as an electronics tech with oscilloscopes and test gear.

We were in monsoon season. There were 4 or 5 major downpours per day. We lived in a Quonset hut on posts with flaps that opened up and screens to let the air in for circulation. Everything was on the water. The walkways were like boat piers. We had house girls that worked during the day only and went home at night. The flight line and the runway were under water. The aircraft looked like speed boats taking off and landing.

The RF-4C aircraft started arriving from the 67th at Mountain Home AFB and the 4411th TARC at Shaw AFB. Sgt. George Gallo was the chief of the IR shop, he was an electronics instructor from Lowry AFB ATC and very muscular. He was a good supervisor. Sgt. David K. Williams was night shift supervisor. I was working swing shift some days and some nights depending on the requirements. Then the new requirements came down. The pilots would fly IR sorties every hour all night long as well as day shift. We did not have the manpower to cover this. It would mean working much longer hours.

Then something happened, we got 2 tech reps to help. Charlie Haines from McDonald Douglas and Clyde Allen from Texas Instruments for the Texas Instruments AAS-18 Infrared System. Both were a Godsend. Charlie knew all the sensors and could explain everything to the pilots. Clyde liked to do a complete preflight with the flight line analyzer with a Wolverine Power Unit and an air conditioner hooked up for cooling. Otherwise we would have to have the engines running to get cool air for the IR system. We did a complete preflight for every IR mission at night. We had the best results and highest percentage in the Air Force. The marines came to our base from Da Nang to learn why and to understand what we did. We were that good.

Figure 7.2 Author Steve Weber in field shop.
Adjusting IR recorder sweep circuits with scopes.

But then something happened. Sgt George Gallo was riding in a jeep with a captain friend and a cement truck ran into his jeep. It could have been a mortar. We didn't know. They were pouring concrete tarmac and runway all the time. We had a metal clad runway that was under water. I don't know how the pilots found it at night let alone the day time. We received word that the captain was killed and Sgt. Gallo was seriously injured and was air evacuated to Tachakawa AFB in Japan, because it was believed to be a head injury with blood all over.

It all happened so fast. David K. Williams became NCOIC of the IR Shop day shift and I became night shift supervisor. We were flying IR night missions every hour all night long. We worked night shift from 5pm in the evening to 7am in the morning, 14 hours a day, 7 days a week. We were short of help. It was exhausting but we had the most successful IR missions in the air force because we did a full preflight of the complex IR System right before the start of every mission.

I was promoted to Airman First Class, Sergeant along with Jim Albright, Arthur Bernier and Charles Gonzalez. We all went to Tech school together. Chuck Gonzalez won the radar manual at tech school. Chuck liked sky diving and competition parachuting. Jim Albright bought a motorcycle a Honda 500 and came to Mountain Home with a pickup truck for his motorcycle. Arthur Bernier was a good friend of my roommate Dave Asman, from Columbus, Ohio, at Shaw AFB, and liked music and driving around in Dave Asman's red convertible. Later Bernie wrote me a letter and explained that Dave Asman, who was sent to our sister flight at Ton Son Neut came back as a Hycon Tech Rep and was killed in the Saigon Tet Offensive in 1968 or 1969.

Thanks to Clyde Allen, our Air Force Technical Representative from Texas Instruments, we had the highest IR Success rate in the Air Force. The pilots were risking their lives flying IR Recon missions over Hanoi, North Vietnam at night. Everything had to work perfectly to get the best results possible. We had lost several RF-4C aircraft to SAM's over North Vietnam. We were flying an IR sortie every hour all through the night. For each aircraft we had the crew chief, a Wolverine or MD3 power unit for aircraft power, the IR flight line analyzer, an air conditioning unit, a helium tank with pressure gauges.

We would do a physical check of the system and insure the helium pressure was up to standard. We would insure the magazine was uploaded with 500 feet of film as required. We would go through the preflight checklist and then power up the IR system from the cockpit control panel. We checked all the appropriate signals and waveforms with the IR flight line analyzer. After completion we would write up the preflight in the aircraft forms. We did this for every single RF-4C IR night mission.

We had the highest IR mission success rate in the Air Force. The Marines at Da Nang had RF-4C aircraft and they sent some of their NCO's to our base to see what our secret to success was. My night shift sidekick Gary Curtis had a fiancée back home in Colorado who bought a 5 acre property that really inspired Gary to greatness. We worked so hard on night shift. I remember one time during the monsoons we drove our shop truck across the flight line to midnight chow instead of going around the base to the chow hall. Well it got stuck in over a foot of water. Gary waded towards another vehicle and asked for help. Well we got some help to get out and we also got chewed out for taking a short cut across the runway for midnight chow. I don't know if we ever got to midnight chow that night. Day shift really had it easy in comparison.

Sometimes, the Air Police would stop by our shop and ask if we wanted a ride over to midnight chow. We had to hurry because of the flight schedules every hour. Sometimes we would preflight several aircraft so we could go to midnight chow and relax. In addition to the preflights and postflights we had regular maintenance. Day shift may have removed a receiver or a recorder and we had to check it out in the field shop. We might have to purge the helium or adjust the sweep circuits etc. We only had one IR System per aircraft. We did not have many spare parts. We had to replace the parts to the component level. When an aircraft returned shot up, and it would be some time before the parts would arrive, we would cannibalize the aircraft and remove the entire IR system for parts. We had one aircraft come back with 95 hits. It just sat on the tarmac.

When Thanksgiving came they flew frozen turkeys in with stuffing. We had C-130's flying in all the time doing touch and go landings. Maybe the turkeys they flew in thawed out a few times. Anyways, after the turkey dinner on Thanksgiving Day, lots of guys got food poisoning and had their stomachs pumped so they said. Maybe it was the stuffing. We were sleeping during the day and had to go to work that night. We had breakfast for midnight chow and breakfast in the morning. This was one of the few times that I was happy to be working night shift.

Figure 7.3, Gary Curtis, Jim Albright and Steve Weber getting ready for night shift.

Gary Curtis, Jim Albright and Steve Weber standing by our hooches where the monsoon rains had evaporated getting ready to go to work. The air conditioned hooches had corrugated metal flaps with screens that would lift up and shed the monsoon rains. A sand bag bunker is shown. The ground under the hooches was flooded during the monsoon rain season.

Reconnaissance and ENERGY VISUALIZATION.

By Stephen Weber

Part I. Aerial Reconnaissance Systems.

Chapter 8. The Bob Hope Show in Vietnam.

After the Thanksgiving turkey fiasco, we started wondering about Christmas. Then we heard a rumor about the Bob Hope show. Would the Bob Hope show come to our remote Airbase? As it turns out, the Bob Hope show was going to Aircraft Carriers, to large bases like Ton Son Nhut, Da Nang, Camron Bay, Okinawa, Guam, Korat, Takhli, etc. Udorn was an Air America base and the very last stop. But, the Bob Hope show made it to Udorn with comedian Phyllis Diller and Les Brown and his band of renown. Bob Hope walked around the stage with his golf club in one hand and Miss World from India, Rita Faria in the other hand.

The most unusual thing was that Bob Hope's jokes were on big sheets of paper on a huge easel. After each joke, that sheet of paper would be ripped down to reveal the next joke. The Bob Hope show was the highlight of our tour. Anita Bryant and Vic Damon sang patriotic songs and sex kitten Joey Heatherton was a great dancer. Naturally they took aerial photos of the show and one with the whole cast comes up front in a line for the final bow to thank the audience. Bob Hope's one liners were endlessly fantastic. Everyone was laughing. Bob Hope had the audience thank his staff and entertainers.

By the time the Bob Hope show came to our base before Christmas, we had lost a half dozen aircraft. Some of the pilots ejected and were rescued by our Jolly Green Giants or they were picked up by the North Vietnamese. A propeller aircraft would fly to the downed site right away, establish communication and smoke out the area before the rescue operations. One of our airmen was celebrated at the Bob Hope show for his daring rescues from a helicopter. Not only did Airman Franklin Stevenson jump off the helicopter and rescue the downed pilots, but he also wrote a song about it, "Ballad of the Jolly Greens." He also played and sang the song.

Figure 8.1 The Bob Hope Christmas Show, December 1966. Bob Hope and Miss India, Rita Faria, in the 1966 Bob Hope USO Show at Udorn RTAFB

Figure 8.2 The Bob Hope Christmas Show, December 1966. Comedian Phyllis Diller with the 1966 Bob Hope USO Show.

Figure 8.3, The Bob Hope Christmas Show, December 1966. Sexy Joey Heatherton sings and dances at the 1966 Bob Hope USO Show.

Figure 8.4, The Bob Hope Christmas Show, December 1966. Singer Vic Damon sings with the Les Brown Band.

Figure 8.5, The Bob Hope Christmas Show, Anita Bryant. Patriotic Singer Anita Bryant sings with the Les Brown Band of Renown at the 1966 Bob Hope USO Show at Air America.

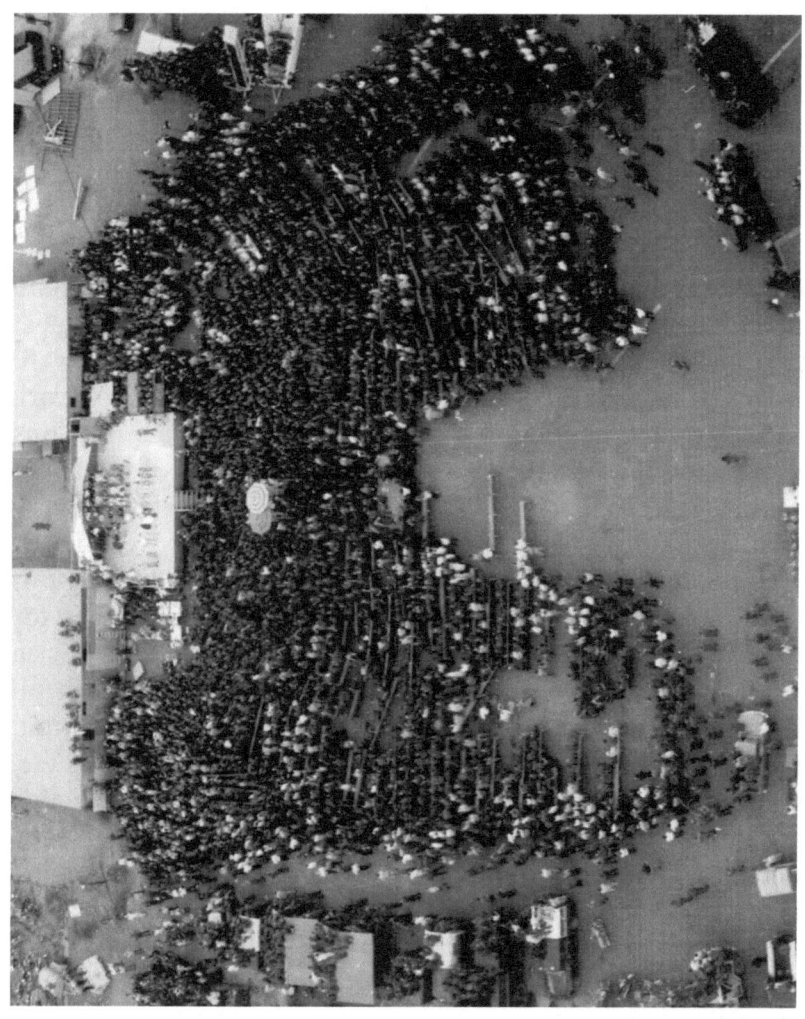

Figure 8.6, The Bob Hope Christmas Show, December 1966. Aerial Photo of the 1966 Bob Hope USO Show finale with all performers on stage at Air America HQ Udorn RTAFB.

Reconnaissance and ENERGY VISUALIZATION.

By Stephen Weber

Part I. Aerial Reconnaissance Systems.

Chapter 9. Electronic Countermeasures ECM.

After the arrival of several RF-4C Phantom II Aircraft around Labor Day 1966, it seemed that the mission was the most important thing. But actually the pilots and the aircraft were the most important. That is why everything had to work properly on the aircraft. We did not know where the missions were going. We assumed they were doing recon of the SAM sites in Hanoi or Haiphong or the Ho Chi Minh trail. Every morning the results of the day and night missions were flown to Ton San Neut Air Force Base in Saigon, Vietnam for PI or Photo Intelligence and then new mission would be planned for the next day and night. When we first got to our Air America Base it was underwater from the monsoons. We shared a metal clad runway with Air America. Air America flew all the old unmarked propeller aircraft like C-123's, C-119's, C-47's etc. We did not know where there missions went either. We also had the Pony Express and the Jolly Green Giants rescue helicopters. Once the monsoons ended and it stopped raining in September and the water covering the runway evaporated, they started pouring concrete runways and C-130's started doing touch and goes. Our RF-4C recon aircraft started flying day and night sorties.

Our recon pilots were well trained at the Tactical Air Reconnaissance Center at Shaw AFB. Even though, a few aircraft were shot down every month. Either during the day or at night, no one said. There were 2 pilots per aircraft. The pilot sat in front and the radar/ photo systems operator PSO sat in the rear. There were 2 canopies. In the day there was ground fire, ak-ak and surface to air missiles. At night there were SAM missiles. I don't know how our pilots found their way home at night, everything was dark. It was scary. Of course they would set the longitude and latitude for our base in their INS Inertial Navigation System before take off and use that information to land safely. But it was dark and you could not see anything. We had the greatest pilots. We lost about a half dozen aircraft. Some of the pilots ejected and were rescued by our Jolly Green Giants and Pony Express helicopters.

Our Air America base was the closest air base to Hanoi and Haiphong. It was a recon and rescue base. If F-105 pilots from Korat or Takhli had a problem they would land at the closest base after a bombing mission. If a bomb was hanging, it could fall off and blow up our runway. In order to rescue the pilots An A1E propeller aircraft would fly to the downed site right away, establish communication and smoke out the area before the rescue operations. We had the Pony Express and the Jolly Green Giants rescue helicopters.

One of our airmen was celebrated at the Bob Hope show for his daring rescues from a helicopter. Airman Franklin D. Stevenson jumped off the helicopter to rescue the downed pilots and won several awards. He wrote and played a song about it, "Ballad of the Jolly Greens."

Saigon (7AF).. TALENTED TROI.. Air Force para-rescue man Airman Second Class Franklin D. Stevson (center) of Ossining, N.Y., is introduced to the audience by Bob Hope and Reita Faria of India (Miss World), during the Bob Hope Christmas show in Southeast Asia. Airman Stevson wrote and sang the "Ballad of the Jolly Greens," about his helicopter - equipped outfit which rescues downed pilots in North Vietnam. The 21-year old airman has been decorated twice for his daring rescues in the communist North. (AIR FORCE PHOTO).

Figure 9.1, The Bob Hope Christmas Show, December 1966. Ballad of Jolly Greens.

The best solution is not to have our aircraft shot down. Perhaps a mission called for flying over or near SAM sites to

54

gather important intelligence information such as the radar uplink signatures for subsequent ECM jamming for future bombing missions. If the radar uplink locked on to our aircraft, a SAM could be launched that could detonate at altitude and scatter shrapnel. We had aircraft land with 95 hits. After the Bob Hope show we received ECM equipment to be installed on our RF-4C aircraft. The IR shop also had the ECM shop in the back. The IR shop was across the runway from Air America with the C-123 aircraft parked in front . We later found out it was operation Ranch Hand that sprayed Agent Orange defoliant in Laos and around our base.

The new ECM equipment was ELRAC made by Hallicrafters in Chicago. Hallicrafters was famous for making high quality HAM amateur radio receivers. Hallicrafters partnered with Lockheed in ECM development. In 1966-67 Northrop purchased Hallicrafters and then partnered with Lockheed in the ECM development for the SR-71 Aircraft. Saving RF-4C Aircraft was an emergency. We helped to install ELRAC in our aircraft. There was a one inch CRT in the cockpit. If a SAM signal was detected by the Hallicrafter receiver an audio alarm signal would be sounded and the pilot would also receive a video signal to tell him when and which way to turn in order to avoid the SAM that was launched.

This saved many RF-4C aircraft from being hit by surface to air missiles. By Christmas it seemed that half of our RF-4C aircraft weren't there. Nobody said anything. Maybe they were sent to our sister flight at Tan Son Neut in Saigon. Maybe the aircraft were hit and the pilots had to eject. We had the Pony Express and Jolly Green Giants helicopter rescue teams. An A1E would be dispatched and smoke out the area. Then a helicopter would land and rescue the pilots. This was done ASAP. Either that or the aircraft and pilot were shot down. This was 1966 and 1967 and there were a lot of our pilots that were POW'S at the Hanoi Hilton in North Vietnam. ELRAC was installed and our aircraft and pilots were safer.

The ELRAC ECM Electronic Counter Measures System made by Hallicrafters, Lockheed and Northrop helped to save our RF-4C aircraft from being shot down by SAM's Surface to Air Missles in missions over North Vietnam. How did they do it? ELINT Electronically Logged Information Transmittal systems on board aircraft like the SR-71 would gather threat information by flying over the SAM sites and record the radar frequencies and SAM site up-link information and bring that information back to be analyzed to determine the type of threats such as SA-2, SA-3, SA-4 etc. Then the ECM systems on board the F-4s, RF-4Cs, F-105s, B-52s, F-104s, F-102s etc would have to be programmed to effectively jam the threats.

One of the first things our Recon aircraft did was to identify and take photo's of the threat environment. Each threat environment is carefully recreated at secret sites in Nevada such as perhaps area 51, which has flight restrictions. Then the ECM systems of various aircraft such as the SR-71 are programmed to jam these threats. Then the aircraft fly ECM missions over these secret sites such as area 51 to test the effectiveness of the jamming algorithms before flying over critical targets such as Hanoi and Haiphong. Of course the SR-71 flies so high and fast over these threats that it is able to tantalize the threats and record the radar signatures on ELINT.

Typically the SR-71 aircraft were in a hanger and would get the OK from Moffit field before flight to be out of view of the Russian satellites. The SR-71 aircraft did not stay at any one base too long. They could be at Beale AFB, AF Plant 42, Clark AB, Udorn RTAFB, Nellis AFB, Edwards AFB etc. The ECM systems had to be programmed to recognize the type of threat before a particular jamming algorithm would be used to jam that particular threat. This was critical because SAM's could be launched to destroy our aircraft. This s why jamming algorithms were verified in ECM test flights over the secret threat installations at secret bases in Nevada.

Reconnaissance and ENERGY VISUALIZATION
By Stephen Weber
Part I. Aerial Reconnaissance Systems.

Chapter 10. R & R.

We were at the 630th Combat Support Group, 432 Reconnaissance Wing for well over 6 months, perhaps 7 or 8 months without an R & R leave, a Rest and Relaxation leave. This was because we were short handed, with our all night, an IR sortie every hour, workload. This was also because our NCOIC of the IR Shop was injured and air evacuated to Tachykawa AFB in Japan. We got several new airman in and I started OJT, On the Job Training for them. I had this idea of visiting Sgt George Gallo in Japan and bringing his foot locker with pictures of his family to him in Japan. So I mentioned it to some of our senior NCO's. They thought it was a good idea. No one knew how bad Gallo was injured or if he would ever recover from a serious head wound.

Figure 10.1, Steve Weber and Chuck Gonzalez on R & R in the old Capital of Kyoto, Japan.

They said that you'll have to get someone to go with you. I asked around and Chuck Gonzalez, who I went to tech school with, thought it was a good idea. Lt. John Rubin from Evanston was probably instrumental in cutting the orders if any. He was more concerned with who was going to take my place. I guess David K. Williams and Gary Curtis would handle the evening shift along with the tech reps. I think we trained some guys but I can't remember who. Perhaps it was some of the guys from day shift photo recon that were used to working the flight line.

Finally we got to go. We had some civilian clothes in a bag. We got Gallo's foot locker and Chuck and I left Udorn on a C-141 to Ton Son Neut in Saigon and then flew to Tachykawa in Japan. Flying over Vietnam at night you could see the clouds being illuminated from all the ground fire below. When we got to Tachykawa we found out where the hospital was and brought Gallo's foot locker to his room. We were so surprised to see him sitting up in bed with a huge cast on his arm. He was happy to see us. He asked us to sign his cast. We explained that everyone thought he had a serious head injury, because his head and face were covered with blood and guts and he was unconscious for some time.

Gallo was a very muscular tech school instructor. He put up his arm when the cement truck ran into his jeep. We explained that the captain he was with was killed. Gallo explained that he knew the captain for a long time. The captain when he was an NCO went back to college to get his degree and receive his commission as an officer. We told Gallo what happened at our base and he said he was coming back. Gallo gave us orders to go to the Navy base, buy what we wanted, and have it sent home. Report back when you're done shopping. We got a very detailed railway map of Tokyo and vicinity from the Tachikawa (East) Library.

The next day we got on the train for Tokyo and headed for the Navy base at Yokohama, the end of the line. There were packers that pushed people on the train like sardines in a can. Then they waved flags to indicate the doors would be closed. Stop after stop people got off and on. All the signs were in Japanese- phonetic characters or Chinese- pictorial and symbolic characters or both. We stood up the entire trip. There was no room to sit down. Finally we got there. We had to show our military ID's to get in the Navy BX. They had everything.

I bought a Pioneer AM/FM stereo receiver and a Nikon 35mm camera with a wide angle and telephoto lens. I thought I bought a Sony tape deck too, but I can't remember. I had them sent home from the Navy BX and later sold them when I was remodeling my house in Glenview. Afterwards we made our way back to the train station at Yokohama, hope we got on the right train and went to Tachyikawa to report back to Sgt George Gallo in the hospital.

Charles Gonzalez had taken up sky diving and did quite a few jumps, when I was sent to Mountain Home AFB to set up the IR shop there. Actually, that was interrupted, he was sent TDY to Korat, Tak Li and Ton Son Nut, Vietnam. I can't remember what Chuck bought at the Navy BX, but I'm sure it was similar to what I got. We both admired the beautifully engraved shotguns there, but they were expensive. Everything was so modern in Tokyo. After being in the military for 3 ½ years, we could not believe how fantastic everything looked.

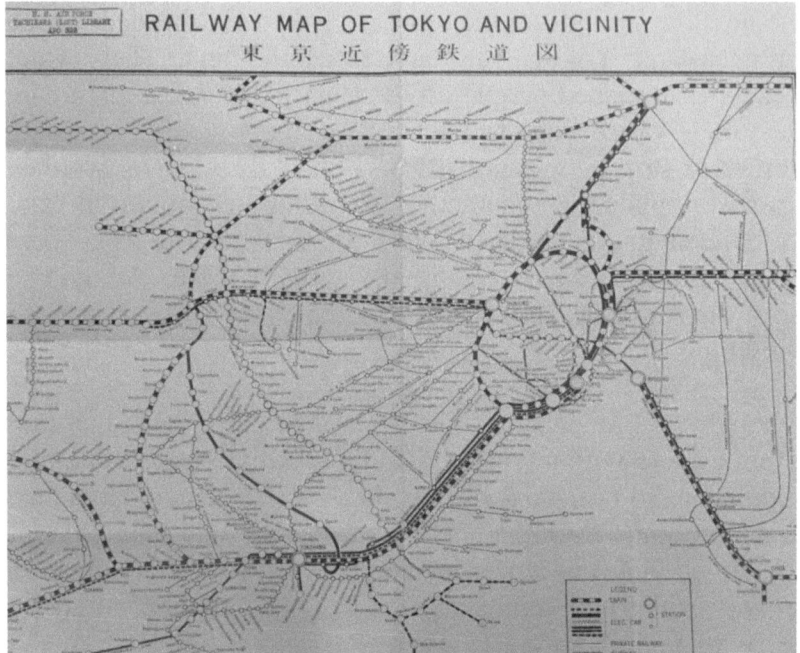

Figure 10.2 Railway map of the Tokyo, Japan and vicinity Railroads from Tachikawa Library.

That evening we visited Sgt Gallo in the hospital. We told him how fantastic the Navy BX was. We told him about the unbelievable number of train stops on the way to the Navy BX. We told him about all the signs in Japenese or Chinese that we could not understand. I'm sure George wanted to visit it but he was in traction at the hospital with a huge cast on his arm. I think that Sgt. George Gallo really wanted to have gone with us to the Navy BX, but he was confined to his hospital bed in traction for his compound fractures. We knew they would take time to heal properly so he sent us.

George gave us our next outing assignment. Another train trip. This time it was the Bullet Train. The New Tokaido Line is high speed and goes from Tokyo past Mount Fuji, the highest peak in Japan, to our destination of Kyoto, the old capital of

Japan. This is a vacation and reservations may be required. George took care of the details. He was having as much fun as us, sending us to the places where he wanted to go to or heard about.

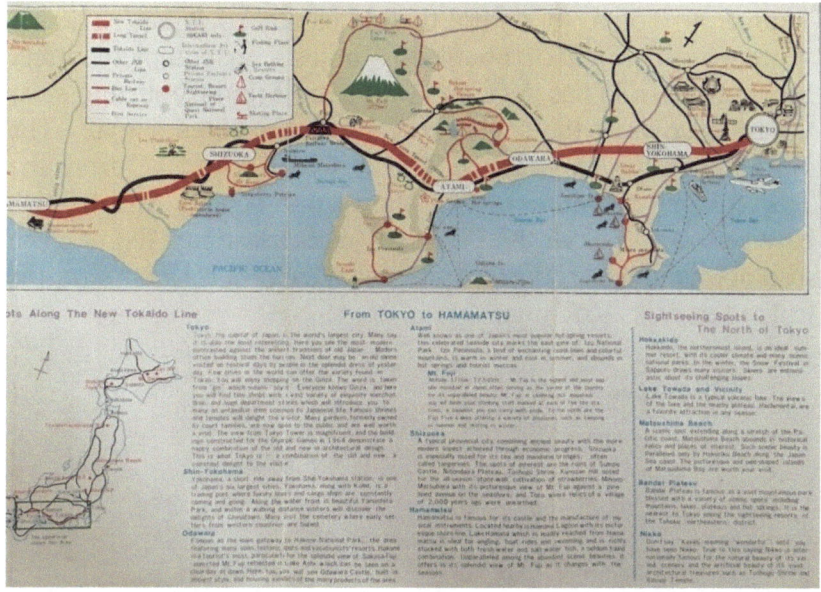

Figure 10.3, The New Tokaido Line Bullet Train map from Tokyo to Osaka, part 1.

The next morning we took the train to the New Tokaido Line station in Tokyo. The Bullet Train rides much higher up like an elevated train on a raised road bed. All the cross streets go through the raised road bed so that the path of the Bullet Train is not disturbed with stop lights. The views out the Bullet Train windows are fantastic or picturesque. We wore baseball hats and carried sports bags. We looked athletic. The people probably thought we were baseball players. As we traveled you could see Mt Fuji majestically out the windows every time you looked. It may have visible for about an hour. Mt. Fuji is 12,500 feet tall and the top was snow packed. Mt. Fuji is a symbol for the country. There are 10 stations for people who want to climb Mt Fuji, five golf courses for playing golf, and five lakes for recreation, camping and skating in winter.

We were passengers on the high speed electric railway Bullet Train from Tokyo to Kyoto, the old capital of Japan. The top speed is 210 km/h or about 131 miles/h in complete safety because the train has ATC or Automatic Train Control to slow the train down if the speed exceeds an indicated signal. The system also has Centralized Traffic Control or CTC from the Tokyo Control Center which shows movements of all trains between the Tokyo and Osaka with flashing lights. The distance is about 300 miles and it takes about 4 hours on the limited express Kodama.

Figure 10.4, The New Tokaido Line Bullet Train map from Tokyo to Osaka, part 2.

You can still see Mt. Fuji from the old provincial city of Shizuoka, known for its mandarin oranges, tea and 2000 year old ruins. Hamamatsu, the next stop, is famous for its musical instruments, castle and views of Mt. Fuji. As we traveled from

Toyohashi to Nagoya, we passed the center of Japan's Automobile Industry. Nagoya is the gateway to Ise-Shima National Park and Nagoya Castle. The next cities are Gifu-Hashima and Maibara, which is the hot springs district and Mt. Ibuki for hiking and skiing. After about 3 hours and 300 miles the Bullet train arrives in Kyoto, the old capitol with its beautiful temples and shrines and three famous festivals which preserve the historical costumes and traditions of the famous old capitol of Kyoto. Even though we had an ice cream bar served by the stewardesses aboard the Bullet Train, we were still hungry. We had lunch at the Kyoto Station Hotel where we made our accommodations to stay. In the afternoon we went on a temple tour and visited Kiyomizu Temple, Sanjusanendo Hall, a 13th century Buddhist temple converted to a museum with 1000 sculptures, and Ryozen Kannon Temple with the Goddess of Mercy and finally Heian Shrine.

That evening we had shredded beef Teriyaki with Teriyaki sauce and chop sticks and beer. We found out that Kyoto has thousands of ancient temples. They are shaped like tall pine trees with their branches sweeping down. Kyoto is also the historical home of the Geisha girls. The next morning we decided to go to a green tea ceremony served by Geisha girls at a famous temple which had many stories each with their own roofs like a pine tree with many boughs. The dried tea leaves were served in a bowl into which was poured boiling hot water. We all sat with our legs folded on the ground. The Geisha girls had beautiful long kimonos, powdered white faces and sat up very gracefully. The green tea ceremony was very formal and very good.

Figure 10.5 Pictured above, Kyoto green tea ceremony with Geisha girls.

We mentioned that we were going on a temple tour and that we were at the Kyoto Station Hotel. The Geisha girls mentioned that tonight was the Silver Night Tour Sukiaki dinner party with Geisha girls. We mentioned that we would be happy to see you there. We went on the Higashi Honganji Temple tour, and on to the Gold Pavilion in a quiet woodland garden, and the Old Imperial Palace, and then the Nijo Castle which had famous interior decorations. After lunch we had the afternoon off and looked at some temples, beautiful rock gardens and the Kyoto Tower which had elevators up to the observatory which is 100 meters above ground.

The Kyoto silver night tour features a sukiyaki Geisha party. Sukiyaki is one of Japans most popular dishes and consists of thin slices of meat, typically beef, tofu which is a soy product and vegetables such as mushrooms and cabbage and soy sauce. Afterwards the formal Geisha girls entertained us with singing and dancing. It was a great evening. Kyoto is a very traditional city.

We were very happy that George Gallo recommended visiting Kyoto the old capitol of Japan.

    The next morning it was time to return to Tokyo to visit the Tokyo tower which at 331 meters was three times higher than the Kyoto Tower and then to Tachikawa AFB to visit George Gallo and tell him of our adventures in Kyoto before returning on a C-141 aircraft Ton Son Neut and then back to work. It was a successful R & R to see the miracle recovery of Sgt Gallo.

    Everyone thought he had a serious head injury when he was air evacuated to Tachikawa Hospital after the collision of his jeep with a cement truck. Sgt Gallo was very muscular for an electronics instructor and he put up his brawny arm to shield himself from the collision and his arm was badly broken. Sgt Gallo was happy to hear about Mt. Fuji and how you could see it for about an hour looking out the Bullet Train window. He had been confined to his hospital bed for months and he enjoyed our travels. Sgt Gallo taught electronics tech school in Lowry AFB in Denver, Colorado so he was familiar with mountains, especially the big ski mountains like Aspen, Breckenridge and Rocky Mountain National Park. George Gallo said he wanted to return and get back to work. We mentioned that our R & R was almost over and we had to go home.

We flew back on a C-141 from Tachikawa AFB to Ton Son Neut. Flying over Vietnam at night you could see the clouds illuminated from the rockets and mortars. We had an adventure that would last a lifetime. Landing in Saigon we bounced 20 feet in the air from the steep descent. After refueling we took off for Udorn RTAFB, the International Headquarters for Air America. Air America flew the unmarked propeller aircraft on secret missions. We told the guys about our trip and how Sgt Gallo was doing. They were jealous but happy to hear that Sgt. Gallo was OK.

Reconnaissance and ENERGY VISUALIZATION.

By Stephen Weber

Part I. Aerial Reconnaissance Systems.

Chapter 11. Teaching OJT, TDY, Agent Orange and FIGMO.

After a week for our R & R to visit injured Sgt Gallo, we still had the highest IR mission success rate in the Air Force thanks to Sgt David K. Williams who took over command when Sgt Gallo was injured. The marines came from Da Nang, they had RF-4C recon aircraft too, to see how we had such great success. We showed them how we would pre-flight each mission with an MD-3 or Wolverine Power Unit, an Air Conditioner and our flight line analyzer. After we purged the cryogenerator with helium so that it would cool down towards absolute zero, we powered up the IR system. One of us would be in the cockpit to run the system, while the other one would use the flight line analyzer to insure everything was working properly. It not only had to work, but it had to produce a high quality image suitable for photo intelligence and reconnaissance.

Last summer, we had metal clad runways that were under water with the monsoons. It was very difficult for our pilots. They were fantastic. We lost several aircraft due to SAM's. Our new concrete runways have been under construction for a long time. We were the closest air base to Hanoi and Haiphong. F-4 and F-105 aircraft from Korat and Takhli would now land at our base if there was a problem such as bombs hanging up and tear up our runway. When this happened we would have to go TDY with our aircraft for several days or a week to continue the IR recon missions as required. Whereas before, with our metal clad runway under water, there was very little interest. When this happened we went on a few TDY's (temporary duty) to other air bases.

We knew that Air America was conducting spraying operations with their C-123 Providers that were parked across the tarmac from our IR/ECM flight line shop. We did not know that Agent Orange was so dangerous and caused so many cancer related problems down the line in years to come. They usually sprayed around the perimeter of our large airbase. We thought they were spraying for mosquitoes whereas, it was a dangerous defoliant that was being sprayed. We later found out it was Agent Orange. Our base was the international headquarters for Air America.

The air base was an old WWII base. We were the 630th combat support group. We had the Jolly Green Giants and Pony Express for rescue operations and the 432nd Tactical Recon Wing with a double wing of RF-4C Phantom II aircraft and RF-101 Voodoo aircraft for day reconnaissance. We had F-102 Delta Dagger interceptors and F-104 aircraft because we were so close to North Vietnam and China. We had many other aircraft that would make special appearances such as C-130s that would do touch and go landings. As shown on the airbase map, there are 2 runways not including the old WWII Japanese runway which is at right angles to runway 12 at 120 degrees and runway 30 which is opposed at 300 degrees.

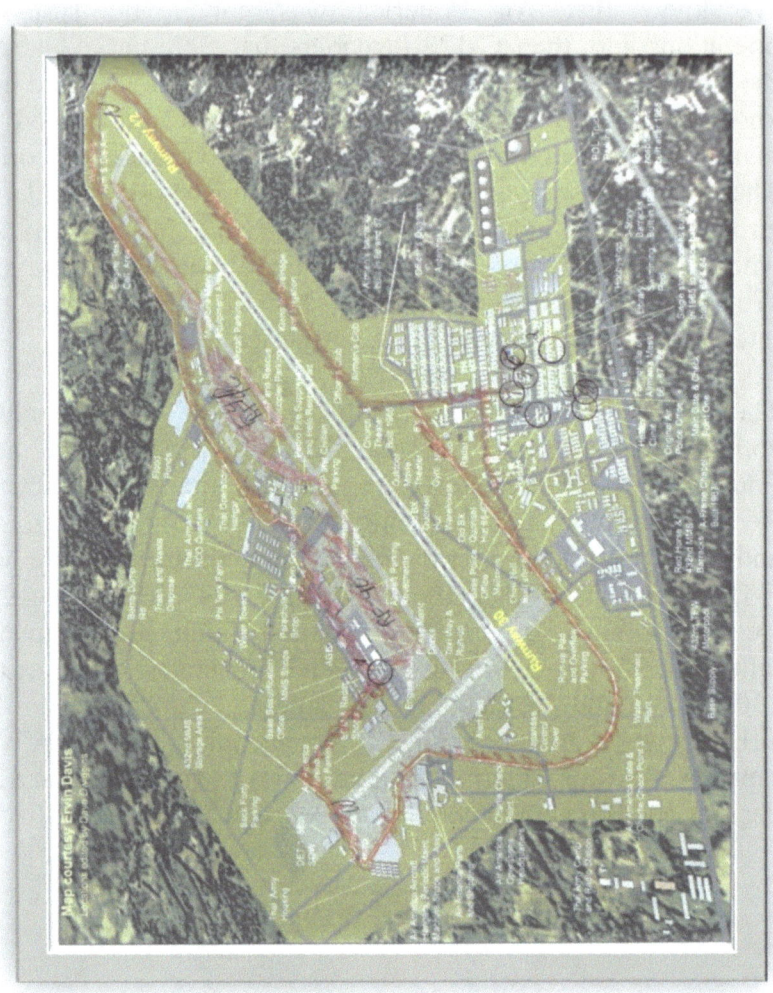

Figure 11.1 Udorn RTAFB runway map.

Air America International Headquarters is a large complex of hangers that face the old runway used as a tarmac that runs east and west that was used by Japan in WWII. The main runway runs north and south and the 630th Combat Support Group aircraft including the 432nd Recon Wing are on the west side of the runway facing east. Runway 300 and runway 12 is an active runway serving both Air America and the 432nd

Tactical Reconnaissance Wing RF-4C's as well as the Jolly Green Giants, the Pony Express, and the A1E's that were used to smoke out the area of a downed aircraft so that the Jolly Green helicopters could rescue the flight crew.

The bus route on the map is marked in red and brings airmen from the chow hall to work at the various flight line shops. The bus brings civilian workers from off base to work on base. The bus route goes around the perimeter of the air base flight line and then off base and into the town. In order to go from base housing units to work on the flight line, the bus goes off base in town.

The Quonset hut hooches are built on posts so that when it rains in the monsoon season the units are not under water. The weather gets hot in April, May and June and the bugs come out in droves. Then the monsoons start in June, July and August. The rains are so heavy that the bugs get washed away. It rains 4 or 5 times during the day and you can see the water evaporating up as water vapor. Then in October, November, December, January and February the weather was dryer.

After we got back from our R & R, we concentrated on setting up "On the Job" training to insure the rigorous quality control standards that we established were followed by the newer airmen that arrived from tech school. We had the highest mission success rate in the air force. Our AF pilots risked their lives on every sortie flying over Hanoi in their unarmed recon aircraft. All critical systems had to work perfectly. We started with a double flight of RF-4C's and by Christmas we were down to one flight. We don't know what happened to the aircraft or the pilots. We had the Jolly Green Giants and the Pony Express helicopters to rescue the pilots. We had the A1E aircraft to smoke out the aircraft. After we installed ELRAC ECM systems we did not lose aircraft. We went on TDY's.

Figure 11.2 RF-4C Phantom II Aircraft Flight Line with Wolverine power unit & air conditioner.

Figure 11.3 RF-4C Aircraft on flight line and a C-123.

Figure 11.4, Rf-4C, A1E, and F-104 salvage area for parts.

Aircraft salvage area, while waiting for parts on the tarmac, sometimes parts or systems would be cannibalized to keep other aircraft operational. We had some aircraft fly back with dozens of bullet holes. The RF-4C has amazing redundant hydraulic lines, neumatic lines and mechanical controls.

Figures 11.5. TDY's to Saigon RVN, Da Nang, and Ubon etc.

Figure 11.6 C-123 Provider Aircraft in flight, AF photo.

Figure 11.7 C-123 Provider Aircraft spraying Agent Orange.

The C-123 Provider aircraft were used extensively for spraying Agent Orange herbicide in Vietnam, Thailand and

Laos by operation Ranch Hand and Air America at Udorn RTAFB. The amount of Agent Orange sprayed was unbelievable as evidenced in numerous reports. "From 1962 to 1971 almost 11 million gallons of Agent Orange were sprayed in Vietnam, primarily through and aerial spray program codenamed Operation Ranch Hand." The aircraft used most commonly for spraying Agent Orange were C-123 Providers.

The C-123 Aircraft would be parked at Air America across the tarmac from our IR/ ECM/ Radar shop. During the monsoons the rains would wash the Agent Orange residue from the planes onto puddles on the ground and we would walk through that on the way to preflight RF-4C aircraft on the tarmac. The 55 gallon drums were around the Air America hangers. Some three million American veterans served in South East Asia.

Agent Orange is a dangerous herbicide and defoliant that affects your genes in addition to killing the green trees and green foliage that it is sprayed on. Agent Orange not only adversely affected all the people that came in contact with the dioxins, but also killed the all green trees and plants it was sprayed on. Agent Orange kills green foliage. Thus Agent Orange contributed to global warming.

The green leaves on the trees absorb carbon dioxide and give off oxygen to fight global warming. Photosynthesis is a natural process needed to combat global warming.

FIGMO. Finally I got my orders. After our one year tour of duty, we were sent back to the States. After the heat in South East Asia and Vietnam, we thought it would be nice to be stationed up north. I received orders for Westover AFB. Bernier got stationed at Pease AFB. Gonzales got orders for Mountain Home AFB where Albright and I came from. Albright received orders for Plattsburgh AFB in upstate New

York. After we were there for a while we decided to go to Expo 67 together in Montreal, Canada for a day. It was fun.

After 4 years in the Air Force, Charles Gonzalez studied to get his first Class FCC license, with a Radar endorsement and get a job in the television industry. Bernier bought an organ and was taking music lessons. I worked at Zenith Radio for a year and went back to school on the GI Bill and studied electronics and physics at NIU. Albright had his motorcycle and truck.

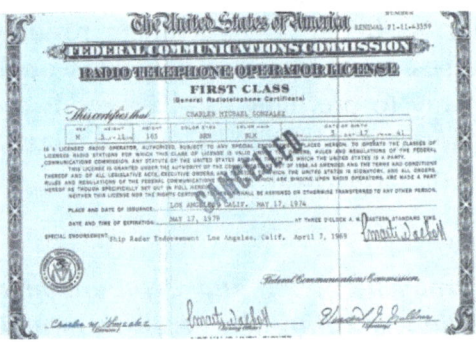

Figure 11.8 First Class FCC Radio Telephone License.

After working at Zenith in 1968, I met 3 great veterans at NIU in 1969 at NIU while majoring in Electrical Engineering. My roommate Chuck Hillebrand was in the Army, We solved the electronics problems on computer paper. Robert Brown was an Air Force radar expert stationed up in Maine. He studied for his First Class FCC license and became a television station manager in Rockford before coming to NIU. Jack Pope was a Navy ship board radar vet and earned college money crop dusting during the summer. We would rent an aircraft and fly. Jack built an aircraft and flew it to Oshkosh. Chuck bought 900 acres of farm land in northwest Illinois. The leader of our group R.W. Brown built an electronics business and passed last year before the pandemic. I bought an old cottage and an old boat in Lake Geneva and we go up there to mow the grass.

Reconnaissance and ENERGY VISUALIZATION.

By Stephen Weber

Part I. Aerial Reconnaissance Acknowledgements:

Figure 1.1 Vietnam, A History by Stanley Karnow.
Figure 1.2 Saturday Evening Post, December 14, 1963. Cover
Figure 1.3 Saturday Evening Post, December 14, 1963 P 24,25
Figure 1.4, Saturday Evening Post, December 14, 1963. pg.29
Figure 2.1 RF-4C Phantom II, Wright Patterson AFB Museum
Figure 2.2 RF-4C Phantom II, flight line photo patch decal.
Figure 2.3 RF-101 Reconnaissance Aircraft, AF photo.
Figure 2.4, Lockheed SR-71, AF Museum with Steve Weber.
Figure 2.5 Televised Electronics.
Figure 2.6 Televised Active Ckts.
Figure 3.1 RB-66 Reconnaissance Bomber.
Figure 3.2 TAC AERO Club Piper Cub.
Figure 3.3 TAC AERO Club Cessna 150.
Figure 3.4 RF-101 Voodoo Recon Aircraft.
Figure 3.5 KS-72 Forward Oblique 5 inch film.
Figure 3.6 Field Shop Gonzalez and LaRegina.
Figure 3.7 McDonald Douglas RF-4C Phantom II Aircraft.
Figure 4.1, IR Reconnaissance System Educational Diagram.
Figure 5.1, 1963 Volvo P-1800 sports car with Steve Weber.
Figure 6.0 a,b. Aerial Photo with ADAS time and date stamp.
Figure 6.1, Author Steve Weber with his 1963 Volvo P1800.
Figure 6.2, 1963 Volvo P1800 in front of author's home.
Figure 7.1 Author Steve Weber working as an electronics tech.
Figure 7.2 Author Steve Weber in field shop recorder sweeps.
Figure 7.3, Gary Curtis, Jim Albright and Steve Weber.
Figure 8.1 The Bob Hope Christmas Show, December 1966.
Figure 8.2 The Bob Hope Christmas Show, Phyllis Diller.
Figure 8.3, The Bob Hope Christmas Show, Joey Heatherton.
Figure 8.4, The Bob Hope Christmas Show, Vic Damon.
Figure 8.5, The Bob Hope Christmas Show, Anita Bryant.
Figure 8.6, The Bob Hope Christmas Show, Aerial Photo.

Figure 9.1, The Bob Hope Christmas Ballad of Jolly Greens.
Figure 10.1, Steve Weber and Chuck Gonzalez on R & R.
Figure 10.2 Railway map of the Tokyo, Japan and vicinity.
Figure 10.3, The New Tokaido Line Bullet Train map part 1.
Figure 10.4, The New Tokaido Line Bullet Train map part 2.
Figure 10.5 Pictured above, Kyoto green tea with Geisha girls.
Figure 11.1 Udorn RTAFB runway map.
Figure 11.2 RF-4C Phantom II Flight Line with Wolverine.
Figure 11.3 RF-4C Aircraft on flight line after pre flight.
Figure 11.4, Rf-4C, A1E, and F-104 salvage area.
Figure 11.5 TDY's to Saigon RVN, Da Nang, Ubon, Tahkli.
Figure 11.6 C123 Provider Aircraft in flight.
Figure 11.7 C-123 Provider Aircraft spraying Agent Orange.
Figure 11.8 First Class FCC Radio and Telephone License.

Reconnaissance and ENERGY VISUALIZATION.

by Stephen Weber

Part II. Light Energy, Aviation and Visualization.

Chapter 12. Light Energy.

Light from the sun warms the earth and enables us to see – to visualize everything. The French impressionists went outside to paint the world bathed in light as they saw it. They were painting light. They interpreted light as color. The colors changed dramatically during the day as the sun rose and set. Claude Monet, the leader of the impressionists, painted many canvasses of haystacks, lily pads and poppy fields each at different times during the day. As the sun moved, the light and shadows changed creating a different impression of the scene. This Plein Air painting was a departure from the great classical paintings of the Renaissance, which were done in the studio. The French impressionists painted light energy as particles of color. According to the wave – particle duality theory of light: waves can exhibit particle –like characteristics, and particles can exhibit wave - like characteristics.

How can we prove that light not only enables us to see but also has energy to warm the earth? I had a Great Course "Prove It: The Art of Mathematical Argument." Light from the sun is a small part of the electromagnetic spectrum. The temperature of each color passing through a prism including the infrared portion was measured and the red portion of the spectrum had a higher temperature. Heat is a measure of energy. Light can be focused with a magnification lens to a point to start a paper fire.

Light is energy as evidenced by Einstein's Nobel Prize for the Photo Electric Effect. When light energy shines on a photocell or solar panel, the light energy is changed to an electric current. Solar panel arrays are used to convert the solar energy into clean electrical energy to save money. The impressionist painters looked to Physics to understand the nature of light, which was thought at that time to travel as particles or packets. The impressionist painters interpreted light as color and painted with dabs, dots, and brush strokes of bright color. Later the dual nature of light was shown to be both particles and waves. We can calculate the light energy. There are two components for the electric E and the magnetic B at 90 degrees to each other for the electro magnetic wave energy density. These are:

Electric energy density = Electric energy/ volume = $\frac{1}{2}\kappa\varepsilon_0 E^2 = \frac{1}{2}\varepsilon_0 E^2$

Magnetic energy density = Magnetic energy/ volume = $\frac{1}{2\mu_0}B^2$

The Total Energy Density = $\frac{\text{Total Energy}}{\text{Volume}} = \mu = \frac{1}{2}\varepsilon_0 E^2 + \frac{1}{2\mu_0}B^2$

Setting the electric Energy Density = Magnetic Energy Density, $\frac{1}{2}\varepsilon_0 E^2 = \frac{1}{2\mu_0}B^2$

Given the speed of Light in a vacuum (Maxwell) = c = $\frac{1}{\sqrt{\varepsilon_0 \mu_0}}$ = 3.0 x 10^8 meters/sec.

Then $E^2 = c^2 B^2$, square root of both sides E = cB, then $E_{rms} = \frac{1}{\sqrt{2}}E_0$ and $B_{rms} = \frac{1}{\sqrt{2}}B_0$

The average total energy density of sunlight is: $\mu = \varepsilon_0 E_{rms}^2$, and $B_{rms} = \frac{E_{rms}}{c}$

The sunlight Intensity $S = \frac{\text{Total Energy}}{\text{Time x Area}} = c\mu = \frac{c}{2}\varepsilon_0 E^2 + \frac{c}{2\mu_0}B^2 = c\varepsilon_0 E^2 = \frac{c}{\mu_0}B^2$

Consult Chapter 35 Solar Radiation for a more detailed explanation.

In 1917 Albert Einstein proposed the theory of stimulated emissions when light (radiation) hits a photo cell (matter) that three things could happen: absorption, spontaneous emission or stimulated emission. This was based on the concept of spontaneous emission theory by Neil Bohr 1913 model of the atom that electrons can only be stable at a discrete set of distances from the nucleus. The Bohr model was based on Planck's quantum theory of radiation: $\Delta E = E_2 - E_1 = h\nu$

This was the theory for the LASER which stands for Light Amplification by Stimulated Emission of Radiation with the development of the Ruby Laser over 40 years later in which a flash tube connected to a high voltage power supply was wound around a ruby laser. Each end of the ruby rod had a mirror and one end was less polished. When the power was applied, the flash tube would excite the electrons to move back and forth in spontaneous emission. Then stimulated emission would occur and coherent red light would shine out the less reflective end. The ruby laser was developed at MIT and was developed into an experimental laser for night reconnaissance at Shaw AFB Tactical Air Reconnaissance Center in 1964.

Figure 12.1 "Prove-It, The Art of Mathematical Argument." By Bruce H. Edwards

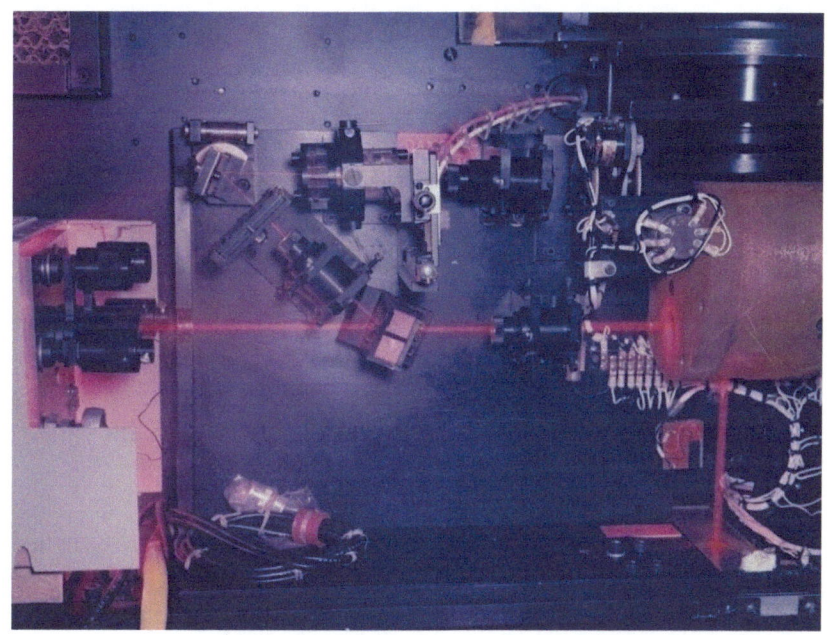

Figure 12.2 The 1964 Ruby Laser used for night reconnaissance flight test evaluation in an RB-66 aircraft at the Tactical Air Reconnaissance Center at Shaw AFB. It was developed by Perkin Elmer out of theoretical Laser research by Charles Townes, Arthur Schawlow and Theodore Maiman.

Reconnaissance and ENERGY VISUALIZATION.

by Stephen Weber

Part II. Light Energy, Aviation and Visualization.

Chapter 13. Prism and Electromagnetic Spectrum.

Approximately 40% of the solar radiation reaches the earth at sea level and has a composition of 38.9% visible light, 54.3% near infrared and 6.8% ultra violet. The IR radiation warms the earth, the UV purifies the air and the visible enables us to see.

Light energy passing through a prism is refracted downward and separated into a rainbow spectrum of colors called dispersion. The colors are red, orange, yellow, green, blue and violet. Red has the longest wavelength 660 nanometers and violet has the shortest wavelength 410 nanometers. These are the colors in the color wheel, electromagnetic spectrum and rainbow. When light illuminates and warms the earth, the objects act as blackbody radiators and emit infrared energy.

The infrared region, discovered by William Herschel in 1800, is actually very large compared to the visible light region. Herschel used a thermometer to measure the temperature of each color of light in the prism and found that violet was the coolest and red was the hottest and beyond the red was even hotter. Herschel called it infrared. The visible light spectrum is a very small portion of the electromagnetic spectrum that extends from low frequency radio waves (AM, FM and microwave), the infrared region, the visible light region, the ultra violet region, x-rays and finally gamma rays.

Table 1. Shows Electromagnetic Spectrum Frequency and Wavelength Range.

Type of waves Range	Frequency (Hz) From	To	Wavelength (m) From	To
Radio	10^4	10^{12}	10^4	10^0
Microwave	10^8	10^{11}	10^0	10^{-3}
Infrared	10^{12}	10^{14}	10^{-3}	10^{-6}
Visible Light	4×10^{14}	7.9×10^{14}	10^{-6}	10^{-6}
Ultraviolet	7.9×10^{14}	10^{16}	10^{-6}	10^{-8}
X-rays	10^{16}	10^{20}	10^{-8}	10^{-11}
Gamma rays	10^{20}	10^{24}	10^{-11}	10^{-16}

The equation: Velocity = Frequency x Wavelength, $v = f \lambda$

In a vacuum v = c, the speed of light = 3×10^8 meters per sec, then $c = f \lambda$ and

solving for the wavelength $\lambda = c/f$, $= 3 \times 10^8 / f$.

The frequency ranges from 10^4 Hz to 10^{24} Hz and the wavelength λ ranges from 10^4 meters to 10^{-24} meters.

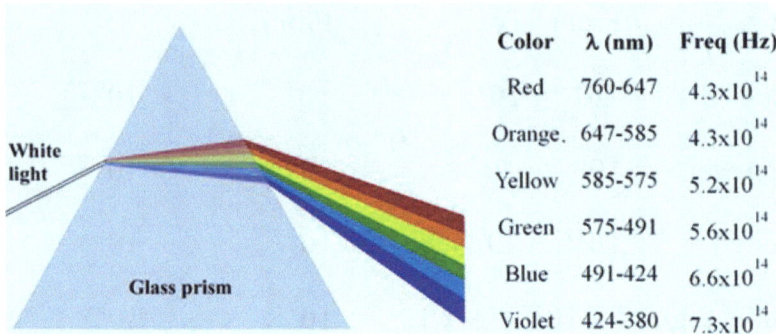

Figure 13.1 Diagram of a Glass Prism and the Colors in the Visible Spectrum.

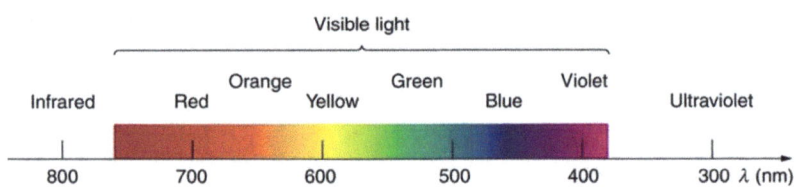

Figure 13.2 Visible Light Electromagnetic Spectrum with IR and UV.

Figure 13.3 Full Electromagnetic Spectrum with Visible Light Breakout.

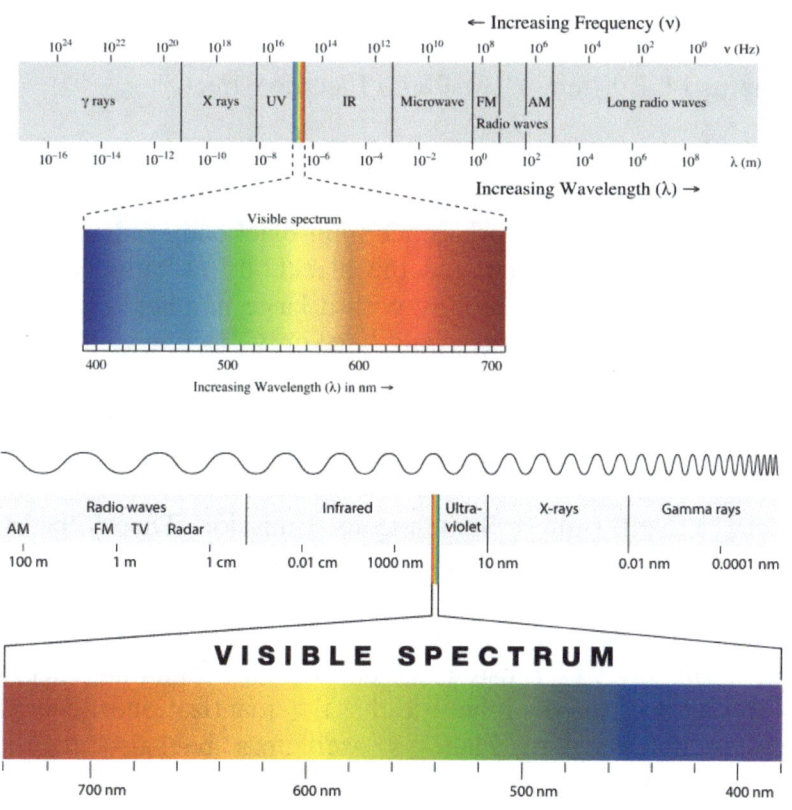

Figure 13.4 Visible Spectrum with Electromagnetic Spectrum in Metric Wavelength.

Reconnaissance and ENERGY VISUALIZATION.

by Stephen Weber

Part II. Light Energy, Aviation and Visualization.

Chapter 14. Einstein – The Photo Electric Effect.

Albert Einstein was awarded the Nobel Prize in Physics in 1921 for his explanation of the Photo Electric Effect in 1905 based upon Planck's work concerning blackbody radiation. In 1900 Max Planck calculated the blackbody radiation curves with a model of a blackbody with a large number of atomic oscillators, and the energy could have only discrete values of n = 0, 1, 2, 3,... Thus the energy is said to be quantized with discrete amounts or packets of energy called photons or photoelectrons in case of the photoelectric effect.

In 1913 Neil Bohr's Spontaneous Emission Theory based upon his model of the atom that electrons can only orbit stably at discrete set of distances from the nucleus of the hydrogen atom. The Bohr model was based on Planck's Theory of Radiation. In 1917 Einstein published "The Quantum Theory of Radiation." In which he said that the quantization of energy applies not only to Planck's oscillators but also to the electromagnetic oscillations of light itself chopping a ray of light into discrete energy packets. Einstein stated that the photon emission can be stimulated by the radiation field. Einstein worked out the equilibrium state between spontaneous absorption, emission and stimulated emission and derived Planck's radiation law. This is the theory for the LASER: absorption, spontaneous emission and stimulated emission.

Electromagnetic waves are composed of particle like photons. Electrons are emitted from a metal surface when light of a high frequency shines on it. When the photoelectrons move towards a positive electrode an electric current is generated. Thus light energy is converted to electrical energy. Light of frequency f is a collection of discrete packets of energy (photons) and the energy E= n h f, where n = 0,1,2,3, and h= Planck's constant = 6.6260755 x 10^{-34} J s.

In 1921 Einstein in his photo electric theory, proposed that light with a frequency f, be regarded as discrete packets or quanta of energy (photons) and each photon with energy E = hf, where h = Planck's constant. When light shines on a metal, a photon can give up it's energy to an electron in the metal.

Total Energy of a particle is the sum of the Kinetic Energy plus the Potential Energy.

Photon Energy = E = $\frac{mc^2}{\left(1-\frac{v^2}{c^2}\right)}$,

and Photon Momentum = p = $\frac{mv}{\sqrt{\left(1-\frac{v^2}{c^2}\right)}}$

Then $\frac{p}{E} = \frac{v}{c^2}$, since v = c for a photon and $\frac{p}{E} = \frac{1}{c}$, then photon momentum p=$\frac{E}{c}$.

The energy of a photon E = hf, and wavelength $\lambda = \frac{c}{f}$, and the momentum p = $\frac{hf}{c} = \frac{h}{\lambda}$.

Einstein used the conservation of energy principle and Planck's model for blackbody radiation for the explanation of the photoelectric effect:

photon energy= max Kinetic Energy + min Work, hf = KE_{max} + W_0 to eject electron.

Einstein's classic equation: $E = \dfrac{mc^2}{\sqrt{1-\left(\frac{v^2}{c^2}\right)}}$, if velocity v = speed of light c, then $E = mc^2$.

Georges Lemaitre, a Belgian catholic priest, mathematician and physics professor proved Einstein's general theory of relatively and proved mathematically that the universe is expanding which later became Hubble's Law. Lemaitre provided an estimate of Hubble's constant. Edwin Hubble and Georges Lemaitre verified that the universe was expanding with the red shift phenomenon when stars are moving away. Hubble proved Lemaitre's theory with his analysis of the Andromeda Galaxy that was thought to be nebulae and was beyond the Milky Way galaxy. Lemaitre's theory became known as the Big Bang Theory. Lemaitre was nominated for the Noble prize in 1954 and 1956. This was beyond Einstein's model of a finitely sized static universe.

Example: Using Einstein's equation E= hf to find the light energy in photons given off by an ordinary 60 watt light bulb. The frequency $f = \dfrac{c}{\lambda}$ and the efficiency of the light bulb is around 2%, then (0.02) (60.0 J/sec) = 1.2 J/sec.

Reconnaissance and ENERGY VISUALIZATION.

by Stephen Weber

Part II. Light Energy, Aviation and Visualization.

Chapter 15. Private Pilot Certificate.

The Federal Aviation Administration Private Pilot Knowledge Exam is required to earn a Private Pilot Certificate. Prospective student pilots may obtain a student pilot certificate through Flight Standards District Office FSDO or a Certified Flight Instructor. The application is vetted by the Transportation Security Administration. Upon approval the applicant can attend an ASA online ground school, Sporty's Pilot Shop program, or a local ground school. While I was working at the University of Illinois at Chicago from 1984 to 1987, I attended the UIC ground school which was taught by Don Gladney who worked for IBM. This was a great evening class which included a Cessna cockpit foldout, a ground school instruction manual and a plastic navigation E-6B guide. The high light was visiting the Flight Service Center FSDO at DuPage County Airport and getting the weather report.

My cousin Junior Johnson was a Flexicore crane operator lifting prefabricated concrete panels typically 8ft W X 30ft L in place to form floors in commercial multi story buildings. Jr. had to imagine a top view while lifting the Flexicore panels vertically and horizontally. So it was a natural for Junior Johnson to attend aviation ground school to get his private pilot certificate. Junior traveled to Chicago to attend a special FAA approved ground school weekend class to pass the FAA written exam in the 1960's. Junior thought the class was great. Then Junior completed his flight training and passed the flight test. After working for years Junior bought his own aircraft. He made a cross country trip to Minnesota and returned safely.

My cousin Vic Isackson went to the University of Illinois and graduated with a degree in Architecture in the mid 1950's and then received a commission in the Air Force. His father Victor was a carpenter foreman for Ragner Benson and very proud of his accomplishment. In 1960 I worked for another Swedish General Contractor Gust K. Newberg as a carpenter building United Airlines Training Center in Elk Grove Village. After OCS, Lt Vic Isakson married his college sweetheart and was stationed in Japan. After four years of soaking up the Japanese architecture. Vic and Sharon moved back to Champaign Urbana and Vic joined the Air Force Reserve at Rantul AFB and started his Architectural practice in Champaign Urbana. Vic started taking flying lessons at the Aero Club and got his Private Pilot's Certificate and Instrument Certificate. Vic rose to the rank of Lt. Colonel before retirement from the Air Force. Of course, Vic Isackson bought his own small air craft.

Another one of my cousins, Roger Anderson, joined the Air Force between the Korean War and the Vietnam War in the mid 1950's. Roger worked as a crew chief at an air base in Texas. After 4 years in Texas in the Air Force, Roger decided to study electronics at Coyne Electronics trade school in Chicago because so much new stuff was happening in electronics, like television. Roger liked the ultralight aircraft perhaps because you did not need to have a private pilot certificate to fly them.

Roger lived next to an ultralight airfield in Rockdale, Wisconsin. The ultralight aircraft were housed in a locked hanger and only flew VFR in fair weather. The pilot was exposed to the weather, since there was no fuselage or windshield to protect you from the elements. Roger also liked model railroading and designed and built his own N scale model railroad on a full 4ft X 8ft plywood layout with 4 engines and digital controls for switching and lots of wires underneath for different hookups and switching systems.

Another one of my cousins, Terry Gayne, was a big crane operator. He worked for Neilson and had the giant crane at Lutheran General Hospital in Park Ridge building additions. He started out as a stock car driver at O'Hare Field when it was an Air Force Reserve base and flight operations were at Midway for propeller aircraft. Terry transitioned to bull dozers and graders when they were building Chicago expressways.

Another one of my cousins Thomas Gorman entered into the military service of our country in 1956 after graduating from Fenwick High School in Oak Park, Illinois. This was after the Korean War Armistice was signed on July 27, 1953 and before Vietnam started in 1961. Comrade Thomas Gorman served in the United States Air Force and attended Air Force Basic Training at Lackland Air Force Base in San Antonio, Texas.

My cousin Thomas Gorman was fortunately stationed at Hickum Air Force Base in Hawaii which was the home of the Pacific Command, PACAF, 15th Wing which was composed of the 15th Operations Group that controlled flight operations, the 15th Maintenance Group that performed Aircraft and Ground Maintenance, the 15th Mission Support which included Logistics and the 15th Medical Group, and finally the Hawaii Air National Guard.

Thomas Gorman worked in the 15th Maintenance Group and held the rank of A1C/ Sergeant and earned the following medals: the National Defense Service Medal (NDSM), the Good Conduct Medal (GCM), and finally he Cold War Citation. Thomas Gorman was Honorably Discharged from the United States Air Force in 1960 and subsequently attended College on the GI Bill in Texas where he studied Engineering. Thomas Gorman served our country with honor and distinction and had an illustrious career in large clean coal power plant construction.

Reconnaissance and ENERGY VISUALIZATION.

by Stephen Weber

Part II. Light Energy, Aviation and Visualization.

Chapter 16. Aviator Friends.

After service in the Navy, Jack Pope returned home and did crop dusting in the LaSalle Peru farming area earning money for the fall semester at Northern Illinois University in DeKalb. Jack was a radar specialist in the Navy and a careful pilot. He may have learned crop dusting from one of the very large farm companies in the area like DelMonte that advertised for pilots to learn crop dusting. Jack had to take medical tests every 2 weeks to make sure that the crop dusting agent did not invade his system. We had several classes together at NIU and in appreciation for help with physics, Jack would rent a plane on a good weather weekend and fly us up to the Playboy Club in Lake Geneva. We would pay the landing fee and fly back to DeKalb Airport.

Years later, Jack would build a hanger and then build his own aircraft from a kit in his spare time and on weekends and then fly his experimental aircraft to Oskosh, Wisconsin for the air show. Jack would have hanger parties around Father's Day with food for the guys to admire his red aircraft in the hanger.

Figure 16.1 Jack, Chuck and Steve at Jack's Hanger Party.

Figure 16.2 Jacks beautiful experimental aircraft.

Figure 16.3 Jack Pope and Steve Weber at the food table.

My neighbor in Lake Geneva, Wisconsin, Aviator Robert, is a Certified Flight Instructor and flies out of Burlington Airport. He started flying 20 years ago with ultralights because you did not need to get your Private Pilots Certificate to fly. However, he decided to attend a ground school in New Lennox, Illinois and then Mike Nevins, M & N, Aviation in Albert Lea, Minnesota where he earned his Private Pilot Certificate and also his multi engine rating in 2005. Aviator Robert continued his flying career and in 2007 returned to M & N Aviation in Minnesota and received his Instrument Certificate and his Commercial Pilot Certificate. In 2008 Aviator Robert went to Alaska to get his Sea Plane Certificate. In 2010 he went back to Alaska and earned his Commercial Sea Plane Certificate.

Aviator Robert is a certified flight instructor and a certified instrument instructor. He bought his own aircraft, a Piper Arrow, and works on his own aircraft as required. He is a very careful and knowledgeable pilot. It is educational and fun to fly with him as we navigate and do touch and go's at some of the local airports or take reconnaissance pictures over the Wisconsin Lakes such as Lake Geneva or Lake Koshkanong.

Figure 16.4 Steve Weber stands by the Piper Arrow.

Figure 16.5 Aviator Robert stands by his Piper Arrow.

Figure 16.6 Aviator Robert's Piper Arrow cockpit instruments.

Figure 16.7 My I-Phone recon photo of a Wisconsin lake.

Figure 16.8 Another I-Phone view of a Wisconsin Lake.

Figure 16.9 Aviator Robert post flight inspection Piper Arrow.

Figure 16.10 Aviator Robert and Piper Arrow back in hanger.

Reconnaissance and ENERGY VISUALIZATION.

by Stephen Weber

Part II. Light Energy, Aviation and Visualization.

Chapter 17. Top View Proposal and Aviator Friends.

Top View Project

Abstract: *A Top View project is proposed to insure Air Traffic Control quality and reliability. The curvature of the earth limits the ground-based radar with a PPI display at the Air Traffic Control Centers to a distance of 200 miles. The basic problem is the transition resolution at the maximum range of the radar as aircraft fly from one ATC center to another, for example: from Cleveland to Chicago or from San Francisco to Los Angeles. The ground radar, like a vector, provides the magnitude or distance and the direction, which has been noted to be off by as much as over 10 miles between zones. In addition, if one zone goes down, there is nothing to provide the continuity between the zones except the airport control towers.*

The Top View project provides the solution to this problem and can be implemented with the current technology at a minimal cost. The Top View project would provide a satellite type view and would use the GPS to provide the exact longitude and latitude of each aircraft. The position of each aircraft as a function of time would be located along with the flight number as a node or vertices on a planar graph on the computer display at each ATC center, with perhaps scrolling and zoom capability. The objective of the Top View project is to improve the resolution of the current system, provide for a dual redundant fault tolerant environment, to insure the quality of the airline transportation system future growth and to facilitate the quality job performance of FAA air traffic controllers.

Stephen E. Weber
EECS 518 Theory of Nets
Dr. W. K. Chen

Ken Petray has been an American Airline pilot for over 20 years. Perhaps he would know if the Top View project was ever implemented. Of course they may not call it a "Top View" system. Perhaps only the air traffic controllers know about these systems and it may be transparent to the pilots. I went through prostate cancer radiation in 2011 around the same time as Ed Petray, Ken's father. VFW post 3579 did an outstanding story of the Petray family on the next page.

WHAT THE VFW AND POST 3579 MEANS TO 3 GENERATIONS OF CURRENT MEMBERS

Our post currently has 3 members who happen to be 3 generations of the same family. It's probably happened before but surely a unique occurrence in this day and age. The Petrays, Ed, Ken, and Kevin are all VFW life members in good standing here at 3579.

Ed Petray has been a post member for over 50 years and served with the 5th regiment of the 1st Marines during the Korean War. He was wounded on 3 February 1953 during a daytime raid labeled "Operation Clambake". Targeting Hills 31 and 31A in the Ungok hill mass, Operation Clambake's success hinged on a tank-artillery feint against enemy positions on Hill 104, Kumgok and Red Hill. The raid took 5 weeks of preparation during which time six rehearsals were conducted. Routes were reconnoitered, mines cleared, fire concentrations plotted as well as practice in casualty evacuation. Shortly before sunrise, the two Marine assault forces, one against each target hill, began their attack. They were met by 3 separate counter attacks from the occupying enemy. Ed was wounded in both legs while repositioning to aid a fallen comrade. The Marine losses during the battle were 14 killed and 91 wounded. Of note, a Marine 2LT named Gerald Murphy was awarded the Medal of Honor for his actions that day.

Ken Petray says that his hero is and always was the Marine that raised him. "Dad never talked about the war but I grew up watching TV shows like "Combat" and "Rat Patrol" and kids would play "Army" in the back yards of Chicago." "I was always a "Marine" when we played and I managed to lose every uniform item my Dad ever had while doing so". The family relocated between O'Hare field and what was the Glenview navy base. "The Saturday we moved in, there were A4s, F4s, and P3s flying low over the house on approach to Glenview and after that, I was going to be a pilot."

Ken deployed in support of the first Gulf war as a C130 Aircraft Commander with the 64th Airlift Squadron attached to the 928th Airlift Wing. He was stationed at SharJah air base in the United Arab Emirates.

Kevin Petray was 3 years old when Ken returned from the Persian Gulf but he does remember sneaking on to the drop zone at Fort McCoy army base while his dad served as the Drop Zone Officer. "We were stopped at the gate the night before and told that children were not allowed on base overnight." "My Dad turned around and returned an hour later with my cousin and me in the trunk. Later that day, an army range control officer saw us on the perimeter of the drop zone and yelled at us for picking up "possible unexploded ordnance" which was actually spent brass from small arms we had found in the dirt." "It's always been airplanes for me as well" says Kevin. "I've been to Oshkosh and have seen the Air and Water show several times. I knew I wanted to fly and the best training in the world is still the U.S. Air Force although I have heard that the Navy runs a little flight school in Florida somewhere."

"Sorry, couldn't help it, Grandpa Ed has another grandchild who is currently at Pensacola going through Marine pilot training and one more about to be commissioned in the Navy headed for submarine duty."

"I have to add that I share my Grandfather and Father's profound love of country, pride in their service and respect for our flag." "They've taught me the importance of family, to appreciate the little things and to take care of others. If you've got dry clothes, a warm bed and a hot meal, it's going to be a good day."

When Kevin returned from his first tour of duty, Ken bought him a life membership to the VFW just as Ed did for him when he returned almost 30 years ago.

Kevin will be deploying at the end of April for his second tour to the Gulf. He is a C130 Aircraft Commander attached to the 757th flying squadron of the 910th Airlift Wing stationed at Youngstown Ohio.

Reconnaissance and ENERGY VISUALIZATION.

by Stephen Weber

Part II. Light Energy, Aviation and Visualization.

Chapter 18. Retina and the Human Eye.

Humans, mammals, and fish have two eyes for fault tolerant redundancy and 180 degree vision. The human eyeball is about 25 mm or 2.5 cm or 1 inch in diameter. Humans and mammals have protective eyelids and eye lashes whereas fish do not. The eye is similar to a camera with a lens, to converge and focus the image, an iris to control the amount of light for the exposure, and the retina or film, to record the image which is inverted from a converging lens; the brain corrects the inversion.

The reflected light of the image enters the eye through the cornea that protects the eye and helps to focus the rays of light through the pupil which controls the amount of light regulated by the iris constricting and dilating, providing exposure control for the light via the muscle surrounding the pupil and then through the lens to focuses the image onto the light sensitive retina or film.

The retina is a light sensitive multilayered tissue and contains the receptor rods and cones and layers of neurons that transfer visual information to the optic nerve to the brain's visual cortex. The cross section diagram of the eye shows this in detail. The job of the retina is to record the image seen by the human eye. The retina is similar to the CCD array in a digital camera, whereas, film in a regular camera is permanently exposed. The retina like the CCD array in a digital camera must be ready for all the subsequent images. In other words the retina and CCD array are reusable.

Figure 18.1 Front view, side view of the eye and eyeball detail

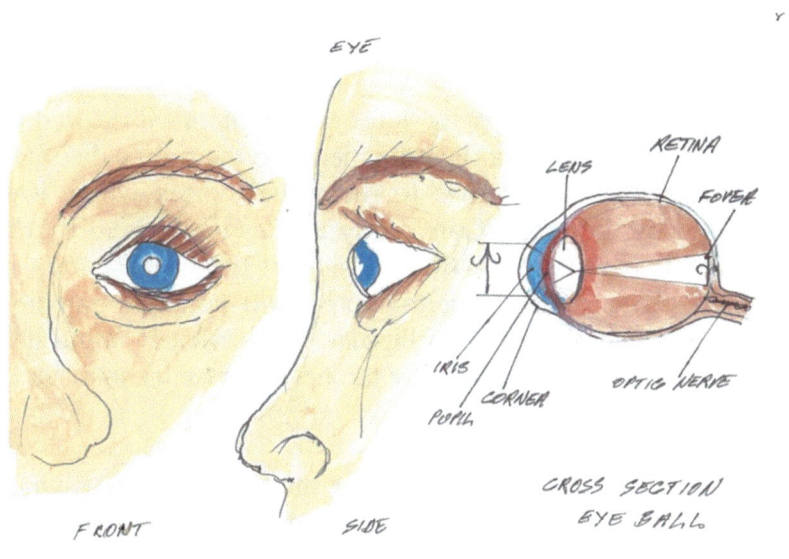

How does that work? The Charge Coupled Device CCD has millions of pixels on a slab of silicon with no wires. Instead the CCD uses channel stops and aluminum bands for the rows and columns. Instead of red blue and green for color the CCD uses an algorithm to average the intensity for color. To get ready for the next picture the CCD employs a shift register and shifts the image up row by row and finally shifts it out to the digital cameras memory chip.

The retina is a light sensitive multi layered tissue with receptor rods and cones with layers of neurons that process the visual information continuously like a movie not just one picture. The detailed diagram of the retina shows light entering the eye and creating photochemical or photo electrical reactions in the rods and cones and bipolar cells which in turn activate ganglion cells whose axons converge to form the optic nerve which transmits information to the visual cortex to the brain's occipital lobe via the thalamus to the brain. The cross section diagram of the retina:

A visual scene viewed by the left eye is temporarily stored by the retina and routed via the optic nerve to the right side of the visual area of the thalamus to the visual cortex where it is processed by the brain. Conversely an image received by the right eye is processed by the left side of the visual cortex in the brain. The retina is multilayered and contains ganglion cells, rods and cones, and bipolar cells. Nearly a million messages can be sent at once by the optic nerve through nearly a million ganglion cells.

Bipolar cells activate the axons of the ganglion cells which converge to form the optic nerve. The rods facilitate low light level vision in twilight in grey, black and white. The cones are clustered near the center of the retina, the fovea, and function in daylight and give fine detail in color. The eye's vision is continuous like a movie. The image data from the retina is continually transferred to the optic nerve and to the visual cortex for visual information processing.

The visual cortex has feature detection neurons which pass information to other areas of the cortex for example face detection in a temporal lobe behind the right ear. As humans we have accumulated a vast visual encyclopedia of faces and places, of tools and cars, of movies and TV programs, of famous paintings and neighborhood houses etc.

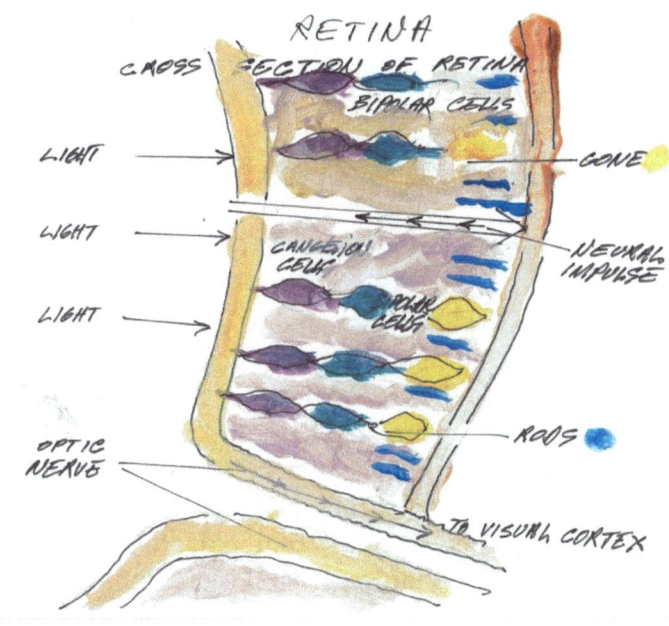

Figure 18.2 Detail of the human eye retina.
Figure 18.3 Parallel Processing and Information Processing.

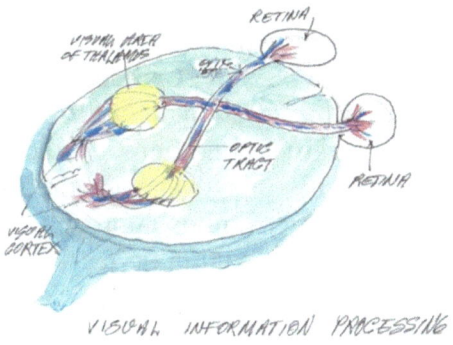

Reconnaissance and ENERGY VISUALIZATION.

by Stephen Weber

Part II. Light Energy, Aviation and Visualization.

Chapter 19. Parallel Processing and Color Vision

The brain uses parallel processing to do several things at once, unlike a single computer which does step by step serial processing. My thesis: "Multiple Asynchronous Parallel Processing for Real Time applications." was initially designed for airborne electronic counter measures and network communication controller applications. The brain works on problems such as form, color, depth and movement simultaneously by dividing a visual scene into subdimensions and then working on each at the same time. Then the more subdimensions, the more elements can be processed in parallel and the task for example, face recognition can be processed at warp speed. But what about the name for that face? Can you remember the name?

 About 35 years ago I had a course in computer vision. One of the problems at that time was auto pilot driving a car with computer navigation. Was it even possible to recognize a stop sign or a stop light or all the other road signs with computer vision with varying light and weather conditions. A technique at that time was to divide the TV image into quadrants and process each quadrant in parallel. Another approach was to divide the TV camera monitor screen with a tic-tac-toe with 9 equal sections with each section processed in parallel and each section looking for a street sign or stop light. Of course at the same time each section would have to look out for other cars and pedestrians too. All this means that the car would be driving very, very slow in order to process all the variables.

How the brain engages in parallel processing is the secret of doing many things at once. Depending on the situation, the brain divides a scene into sub dimensions such as color, movement, depth, and form and works on each element simultaneously to expedite real time processing. The brain reconstructs these elements depending on the situation for decision making.

For example, let's say the person is flying a plane. There are several things that the pilot has to do simultaneously such as scan the sky for other airplanes, scan the instruments to make sure the altitude and direction are appropriate for the desired radio navigation vector.

Another example that requires fast decision making is face recognition; the brain compares the new retinal image to stored visual cortex areas to recognize the image in less than a second. The act of parallel processing depends on the situation and the task at hand. All this happens effortlessly, instantly and continuously to facilitate solving problems.

The cones gathered around the fovia in the retina provide detail and color vision. If we see an object such as an apple it appears to be red. The apple reflects red light. According to Sir Isac Newton the apples color of red is our mental reconstruction.

The Young-Helmholtz trichromatic theory that used 3 colors of light: red, green, and blue to create all the colors. This is how color television works. This is the additive theory of light which adds wavelengths. There is another theory, the subtractive color of light used in mixing paints which uses red, blue and yellow colors and subtracts wavelengths.

The color theory that our eyes use is the Young-Helmholtz that uses the additive theory of red, green, and blue lights. The retina has 3 types of receptors corresponding to these colors.

There are no receptors sensitive to yellow, but when red and green cones are stimulated we see yellow.

The color of an object is influenced by the adjacent colors and the illumination. The brain computers color relative to its environment of the surrounding colors. This is important for artists, architects, interior decoration and clothing. The color wheel is very important to see the relationship of complimentary colors.

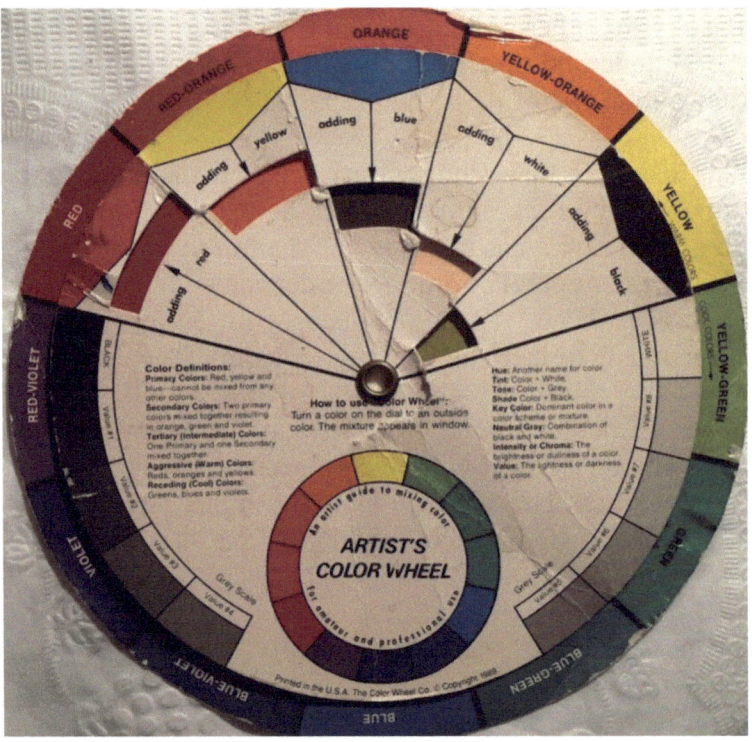

Figure 19.1 Color Vision and the color wheel.

Reconnaissance and ENERGY VISUALIZATION.
by Stephen Weber

Part II. Light Energy, Aviation and Visualization.

Chapter 20. Drawing

An idea is a thought or collection of thoughts that generate in the mind perhaps during brainstorming. An idea can be an opinion, belief or concept. It can be real or abstract. It could be a general idea of how you would design or build something. In this case a sketch or drawing could be made to illustrate the idea. If the idea was a house an architect could draw a floor plan and front elevation of the house in order to distinguish it from millions of other houses. Building and zoning might require a rendering of the proposed house in an historical architectural district for approval.

In historical architectural areas renderings are sometimes required when applying for a building permit in addition to the elevations and floor plans. The renderings are typically done in watercolor or gauche. Sometimes word drawings are made when applying for repair or much simpler repair permits. "One drawing is worth a thousand words". Artists have a sketch book to practice and document their studies. Engineers may have an engineering notebook to record their ideas with drawings. The desire to communicate via a drawing goes back to cave dwellers and the drawings on the walls inside the caves. Drawings and paintings go back thousands of years.

After retiring from teaching 10 years ago, my wife found a newspaper article about painting classes at the local park district. She showed the article to me and said "Why don't you take a class in painting from the park district"? So I paid the tuition, purchased some acrylics and brushes and went to class only to find out they were ringers; professionals who exhibited at the Park Ridge Art League Art Fair. They would look at pictures of dogs or flowers or trees and paint them, rather than go outside and paint from nature. I liked to paint old barns. I even made some frames from rough sawn cedar on the table saw and miter saw. I joined the Park Ridge Art League (PRAL) and exhibited there on Father's Day weekends for several years.

It wasn't until PRAL had a guest artist demonstrate his watercolors, that I was enlightened. After graduating from the Chicago Academy of Fine Art, he got a job as a story board illustrator for an advertising agency. A story board is a series of watercolor sketches or frames for scenes of a story. Story boards were used as paintings of scenes for a movie. In his case he had to create scenes for an advertising movie of a product or service. He had to create something from nothing. He went back to the school and told them that he was not trained to paint something from his imagination and asked what he could do; because he needed the job and was to start in a week. The professor explained to take his sketch book and look at something and then turn away and quickly sketch it from memory. Do this with different subjects each time turning away and sketching it. You will reinforce your memory and your ability to sketch and paint from your imagination. The graduate did this. He became very successful in his job as a story board illustrator and later as a commercial artist.

We visualize so much in our lifetime: ideas, movies, so many TV programs, so many scenes going to work on the train or driving on a vacation. Do we save the whole movie in memory or perhaps just scenes or snapshots? We can remember if we saw the movie before. To what detail can we remember if we had to paint it from memory. How does this visual process work? This is why drawings are essential to accurately communicate ideas in construction, engineering, manufacturing, science and art. The abbreviation STEM stands for science, technology, engineering, and math. STEAM adds art.

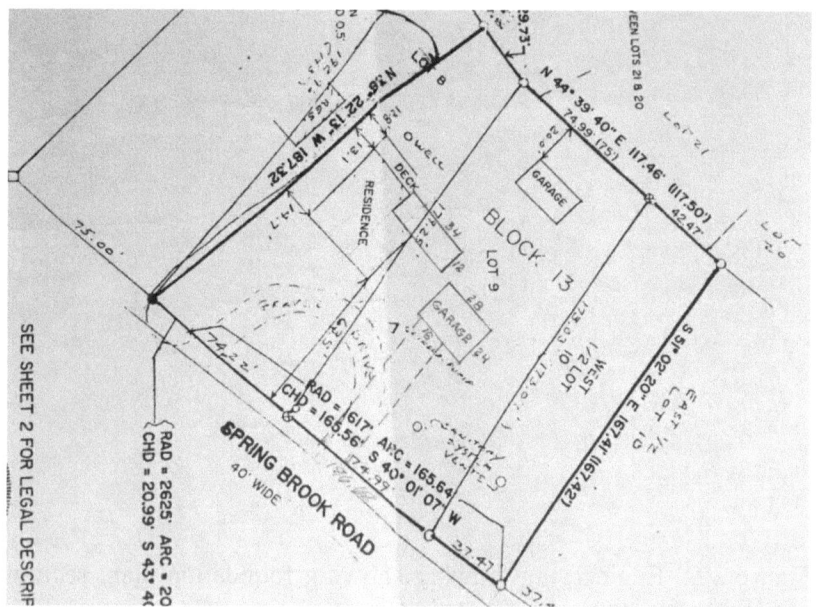

Figure 20.1 Platted survey of our vacation house in Wisconsin.

Another example is our vacation house in Wisconsin. It has a detached 2 ½ car garage that is 26 feet from the house. There is a path to nowhere between the house and the garage going up the hill to the back door. I designed a cathedral beamed living room addition with a front door entry right in the path to nowhere. This design will be used to communicate the idea to my wife, building and zoning, the general contractor, the lumber yard for take-offs (materials costs), the cement foundation and excavation contractor, the carpenter contractor, the electrical contractor, the insulation and drywall contractor, the HVAC - heating, ventilation, and air conditioning contractor, the mason contractor and the interior decorator. I have included several engineering drawings for examples.

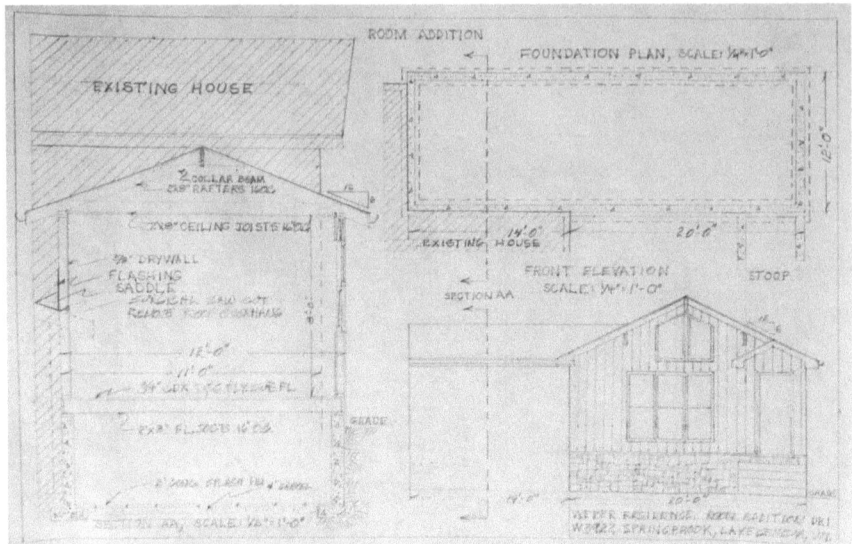

Figure 20.2 Engineering drawing showing foundation plan, section and front elevation.

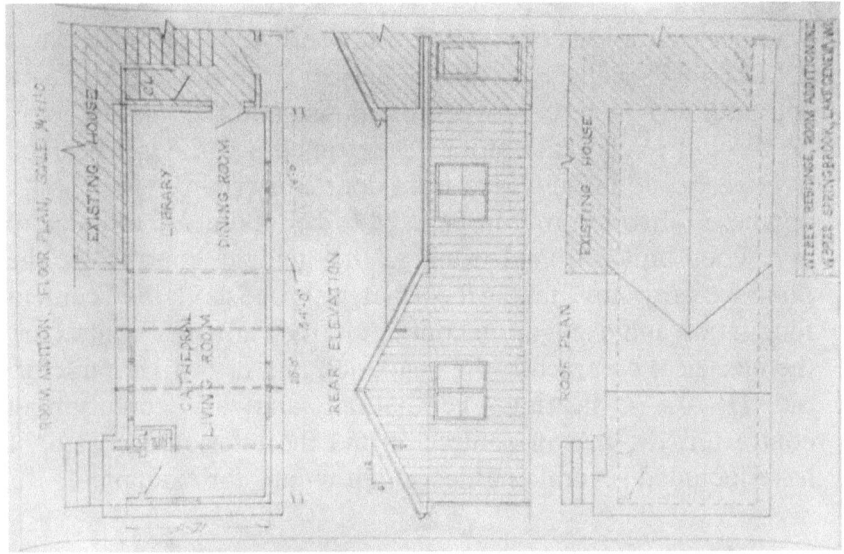

Figure 20.3 Engineering drawing showing floor plan and rear elevation of addition.

Reconnaissance and ENERGY VISUALIZATION.

by Stephen Weber

Part II. Light Energy, Aviation and Visualization.

Chapter 21. Engineering Graphics - AutoCAD.

CAD stands for Computer Aided Design. AutoCAD is a menu driven program that enables us to automate engineering drafting. Instead of using T-squares, triangles, and a drawing board to design an aircraft in 10 years, using AutoCAD is faster, more accurate and can facilitate changes without having to redo everything - all the drawings. The AutoCAD program can run on a Windows workstation, a mainframe or a laptop. The drawings can be printed on a laser printer, a plotter or sent via email to a graphics printing facility. The drawing is displayed on graphics monitor similar to how a word processor displays writing. The monitor takes the place of the drafting paper. Cartesian coordinates are used in AutoCAD with an x, y, and z axis. The dimension units can be scientific, decimal, engineering , architectural, or fractional typically in feet or meters. Drawings can be saved on disk with a file extension of .DWG. For detail the menu has a zoom command.

The drawing editor has a menu bar, a toolbox, toolbars, and a drawing area with a crosshair cursor. Commands can be entered via the keyboard, screen menus, pull down menus, tablet menus and the mouse or digitizer. Keyboard keys are used in AutoCAD include the command line entry key (backspace, ctrl-H, ctrl-X, esc), toggle keys (F1, F2, ctrl-T or F4, ctrl-E or F5, ctrl-D or F6, ctrl-G or F7, ctrl-O or F8, ctrl-B or F9, ctrl-V.
There are other keys: ins, home, up arrow, down arrow, left arrow, right arrow, pg up, and pg dn. There are also dialog boxes and a host of dialog box commands.

If the idea was a house an architect could draw a floor plan and front elevation of the house in order to distinguish it from millions of other houses. Building and zoning might require a rendering of the proposed house in an historical architectural district for approval. A general contractor would use the drawings to determine the materials and costs in order to submit a bid to the prospective owner. Many sets of the blueprint plans may be used by the excavation contractor, the foundation or cement contractor, the carpenter contractor, the electrical contractor, the plumbing contractor, the drywall contractor, the masonry contractor, the painting contractor, the roofing and siding contractor, the interior decorator, the finish carpenter and the building inspectors. In summary, architectural drawings are essential in the construction industry. This is how ideas are communicated.

For example: Our post has proposed to buy a new VFW home instead of renting a corner in a Banquet Hall that may be closing. The proposed building was a doctor's office building. The board of directors wanted a feasibility study. I drove by the proposed building and took a few photographs. Then I made a few sketches with ovals for rooms or functions. Then I made an architectural drawing to 3/16" scale with a front elevation and a floor plan on an 18 in. X 24 in. drawing paper. The front entry opened to the bar area, to the left was the Classical ballroom with a 10 ft-4in ceiling. The building was a 100 ft. long brick ranch with an 8 ft ceiling. To the right of the bar was a cathedral beamed ballroom. Behind the bar was the frame family room style addition, which I made the VFW post meeting room and offices for the commander, quartermaster, and color guard. The problem was the 8 ft ceiling with the center bearing walls, which was partitioned into doctor's offices. I did two section details for the classical and the cathedral ceilings to facilitate understanding.

Of Course, our VFW post wanted beamed open cathedral ceilings designed by a licensed architect because the local village would have a separate agency review the plans in accordance with the building codes. This process took an extra six months just for the architectural drawings to be approved by the Village of Niles

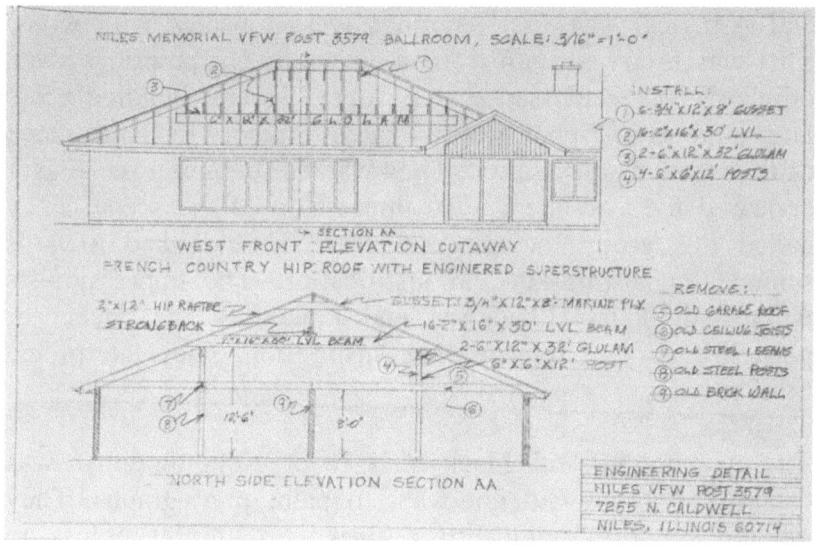

Figure 21.1 Engineering section details of our newly purchased future VFW post in Niles

Reconnaissance and ENERGY VISUALIZATION.

by Stephen Weber

Part II. Light Energy, Aviation and Visualization.

Chapter 22. Painting – Impressionists.

Art history paintings fill museums in every country in Europe and America. The first paintings were mosaics with icons. This was followed with frescos on walls. Oil paintings were popular in the renaissance in Europe. Several hundred years later a group of French artists wanted to record the appearance of the world as they saw it. They wanted to paint a scene as it appeared for a moment – an impression of the scene. They called this group the impressionists. They looked towards science for inspiration and justification. The impressionists painted outdoors (Plein-Air) and looked to the packet or particle theory of light as their inspiration. They interpreted light as color.

The technology of black and white photography was established. They did not want to paint photographs. They painted outdoors with brush strokes of bright colors to represent light. They looked at the prism and the colors of red, orange, yellow, green, blue, and violet. White light contained all these colors. The impressionist paintings of Monet and Renoir were beautiful. This was followed by the postimpressionists, who continued the light as color and tried to add structure to their paintings, like Cezanne and Van Gogh.

Claude Monet was the leader of the movement. Monet exhibited a painting with the name of "Impression." Thereafter the movement became known as the Impressionists. These paintings were painted outdoors with natural lighting. They painted the particles of light as brush strokes, dabs or points of colors. They did not paint outlines or solid colors, because a reflected color was composed of many particles of different colors. There were no lines in nature, but rather the intersection of planes of reflected light as packets of light interpreted as colors.

Claude Monet painted many studies of ordinary subjects at different times of the day like a scientist to observe the different colors and shadows in paintings of water lilies and hay stacks. In fact Monet created paintings of lily ponds that filled an art museum room. He painted these paintings as experiments as he saw it at different times of the day, and not as he knew it to be. Because the light changes very quickly during the day, Monet had to paint many different haystack paintings to capture the different light effects and colors.

Claude Monet was born in 1840 and lived to be 86 years old. His paintings are in the best art museums around the world. The "Musee Marmottan Monet" has a great collection of Monet's Water Lillies and opened in 1940, 100 years after Monet's birth and just 14 years after his death in 1926. Monet's paintings have sold for millions of dollars. Monet was the leader and the scientist of the impressionist painting movement in France. He painted outdoors studying light and color at different times of the day. The camera could record the image in black and white, but the impressionist painter would visualize color.

Figure 22.1 Claude Monet, "Garden at Saint Adresse."

Claude Monet was the leader of the French Impressionist school of art. This landscape/ seascape is the Garden at Saint Adresse with the couple in the forground with a straw hat and a parasol, the couple in the middleground ground between the flags, and the sea as the background with the ships on the horizon line. The center of focus is the white parasol. The patio fence keeps the viewer in the foreground away from the water. The flowers are multicolored all around the patio garden. The masterful control of light in the forground is like a symphony orchestra control of volume so that everything is perfect.

Figure 22.2, Claude Monet, "Poppy Fields near Argenteuil."

Claude Monet was the leader of the French Impressionists. He painted this in 1875 at the age of 35. Monet was born in 1840 and died in 1926. Monet studied light and color while he painted outside "Plein Air" to observe how the light and color changes at different times of the day. In fact, Monet would paint several paintings of the same subject at different times of the day to show how the light and color changed in the scene. Monet did this with hay stacks, cathedrals and lily pads in ponds. Monet studied the effects of light at different times of the day like a scientist and yet his paintings have a fresh child like innocent appearance. The little dabs of color appear fresh. Monet is like a student studying light, color and nature as it appeared to the French Impressionists 100 years ago.

Figure 22.3, Claude Monet, "The Haystack at Giverny."

The Haystack at Giverny by Claude Monet is a beautiful impressionist painting with several planes of depth. The haystack is in the forground. The main focus is the field of orange poppies surrounded by green. The farm houses in the middle along the road with the dark foilage of the trees seperates the forground from the background, which is a bluish gray green. The energy comes from the orange vermillion color of the poppies. Monet painted many pictures of haystacks at different times of the day to study the changing colors.

Figure 22.4, Claude Monet, "Water Lillies."

Water Lillies Green Reflection by Claude Monet is the forground of the painting. It is almost two dimensional except for the reflections and the oval lilly pads floating and getting smaller in the distance. Monet painted many paintings of water lillies. There is a museum of his water lillies on each wall and you sit in the middle and view water lillies.

Reconnaissance and ENERGY VISUALIZATION.

by Stephen Weber

Part II. Light Energy, Aviation and Visualization.

Chapter 23. The Basic Elements of Design.

As we look at the various schools or periods of art, there are basic esthetic elements of design and composition that together create beautiful works of art. These are line, shape, color, composition, balance, pattern, motif, contrast, focal point or center of interest, style, texture and energy. Typically many artists have developed their own style, which is influenced by their period. The impressionistic style which is popular today among traditionalists, include Monet and Renoir.

Many artists such as Monet used foreground, middle ground and background to create depth or perspective vanishing lines to create depth a three dimensional effect on a two dimensional surface. Contemporary artists like Bob Ross use vanishing planes to create depth with pale colors in the background and dark contrast and warm colors in front. Many artists use good composition techniques such as framing on either side with trees. A sky, that is lighter at the horizon and darker at the top and corners. A center of interest or focal point, that is not in the center of the painting. There are other techniques, such as light medium and dark values and shapes of large light areas and smaller darker areas.

In summary, the basic elements of composition and design are used differently by artists in different periods of art history.

Reconnaissance and ENERGY VISUALIZATION.
by Stephen Weber

Part II. Light Energy, Aviation and Visualization.

Chapter 24. Artistic Elements in Painting, Music, and Literature.

The basic aesthetic elements of painting correspond and compare with the other art forms such as music, literature and architecture. For example:

Table 24.1 Artistic Elements in Painting, Music, and Literature.

Painting	Music	Literature
Line	Melody	Plot
Color	Instrument	Character
Shape	Movement	Chapter
Composition	Arrangement	Book
Contrast	Volume	Intensity
Pattern	Chord	Repetition
Texture	beat	detail

These artistic elements are used differently throughout history and differently by different artists to achieve a unique style. The energy in a particular work of art is how these elements are used together.

Reconnaissance and ENERGY VISUALIZATION.

By Stephen Weber

Part II. Light Energy, Aviation and Visualization.

ACKNOWLEDGEMENTS.

Figure 12.1 "Prove-It, The Art of Mathematical Argument." By Bruce H. Edwards.
Figure 12.2 The 1964 Ruby Laser used for night recon.
Table 12.1 Electromagnetic Spectrum Frequency/Wavelength.
Figure 13.1 Diagram of a Glass Prism and the Visible Colors.
Figure 13.2 Visible Light Electromagnetic Spectrum IR/ UV.
Figure 13.3 Full Electromagnetic Spectrum with Visible Light.
Figure 13.4 Visible Spectrum and Electromagnetic Spectrum.
Figure 16.1 Jack, Chuck and Steve at Jack's Hanger Party.
Figure 16.2 Jacks beautiful experimental aircraft.
Figure 16.3 Jack Pope and Steve Weber at the food table.
Figure 16.4 Steve Weber stands by the Piper Arrow.
Figure 16.5 Aviator Robert stands by his Piper Arrow.
Figure 16.6 Aviator Robert's Piper Arrow cockpit instruments.
Figure 16.7 My I-Phone recon photo of a Wisconsin lake.
Figure 16.8 Another I-Phone view of a Wisconsin Lake.
Figure 16.9 Aviator Robert post flight inspection Piper Arrow.
Figure 16.10 Aviator Robert and Piper Arrow back in hanger.
Figure 16.11 What the VFW means to 3 Petray Generations.
Figure 18.1 Front view of eye, side view and eyeball detail.
Figure 18.2 Detail cross section of the human eye.
Figure 18.3 Detail of the human eye retina.
Figure 18.4 Parallel Processing.
Figure 19.1 Color Vision and the color wheel.
Figure 20.1 Platted survey of our vacation house in Wisconsin.
Figure 20.2 Engineering drawing: foundation, section and elevation.
Figure 20.3 Engineering drawing floor plan and rear elevation. addi
Figure 21.1 Engineering section details of our future VFW post in Niles

Reconnaissance and ENERGY VISUALIZATION.

By Stephen Weber

Part II. Light Energy, Aviation and Visualization.

ACKNOWLEDGEMENTS Continued.

Figure 22.1 Claude Monet, "Garden at Saint Adresse."
Figure 22.2, Claude Monet, "Poppy Fields near Argenteuil."
Figure 22.3, Claude Monet, "The Haystack at Giverny."
Figure 22.4, Claude Monet, "Water Lillies."
Table 24.1 Artistic Elements in Painting, Music, and Literature.

Reconnaissance and ENERGY VISUALIZATION.

by Stephen Weber

Part III. Extensions to Our Vision.

Telescopes and Astronomy.

Chapter 25. Galileo and Our Galaxy.

Galileo Galilei, an Italian astronomer, physicist, and engineer was born in Pisa in 1564. He was believed to invent the telescope and was called the father of observational astronomy and the father of modern physics. Early in his life Galileo studied velocity and gravity, relativity, inertia, projectile motion, pendulums, hydrostatic balances, and using the telescope for scientific observations. Galileo's contribution to astronomy include the telescope, the telescopic confirmation of the phases of Venus, the four largest moons of Jupiter, the observation of Saturn's rings, the analysis of sunspots and the theory that the earth and the other planets revolved around the sun. This was called heliocentrism and Copernicanism instead of the old belief that the planets revolved around the earth. For this Galileo was tried by the Roman Inquisition and found guilty of heresy and forced to recant. He spent the rest of his life under house arrest and formulated two new sciences: kinematics and strength of materials.

The irony is that the telescope was invented in Holland in 1608 by Hans Lippershey and Zacharias Jansen spectacle makers who held a lens in one hand and another lens in the other hand and noticed that the church steeple that he was looking at was enlarged. They noted the distance between the lenses and packaged the lenses in a tube and started selling telescopes. In 1609 Galileo heard about this and made a better telescope. He presented his invention in public before the Doge Leonardo Donato and they doubled Galileo's lectureship salary at Padua.

Why were astronomy enthusiasts using spectacle lenses in Holland? Because the Gutenberg bible was printed in 1450, people needed spectacles or reading glasses to read the new Gutenberg bible. There are two types of telescopes: the refracting telescope which uses two lenses and the reflecting telescope which uses 2 mirrors and a lens. When discussing telescopes we are talking about large distances. Distances are measured in both the English and metric system. Large distances use the powers of ten and typically use the metric system in science. Similarly with computers and memory we use the metric system and the powers of ten.

Table 25.1 Computer Memory and Distance.

1 byte = 8 bits, 1 meter = 10^2 centimeters = 10^3 millimeters.

1 kilobyte = 10^3 bytes

1 kilometer = 1000 meters = 10^3 meters

1 megabyte = 10^6 bytes, 1 million meters = 10^6 meters

1 gigabyte = 10^9 bytes, 1 billion meters = 10^9 meters,

1 terabyte = 10^{12} bytes, 1 trillion meters = 10^{12} meters

The speed of light = 186,000 miles / sec = 3 x 10^8 meters /sec

Powers of 10, Distance Measurement

1x 10^1 = 10 millimeters = 1 centimeter
1x 10^2 = 100 centimeters = 1 meter
1x 10^3 = 1000 meters = 1 kilometer
1x 10^6 = 1,000,000 meters = 1 million meters
1x 10^9 = 1,000,000,000 meters = 1 billion meters

1×10^{12} = 1,000,000,000,000 meters = 1 trillion meters

Speed of light (c) = 299,792 km/sec = 3×10^8 meters/ sec.
Light year (ly) = 9.46×10^{12} km = 9.46×10^{15} meters, 1 par sec (pc) = 3.26 light years.

Table 25.2 Exponent Operations:

$\frac{1}{10^n} = 10^{-n}$, $10^n \times 10^m = 10^{n+m}$, $\frac{10^n}{10^m} = 10^{n-m}$

Telescope = Large distances, positive exponents powers of ten.

Microscope= Small distances, negative exponents powers of ten.

1 millimeter = $\frac{1}{10^3}$ $meter$ = 10^{-3} meter.
1 micrometer = $\frac{1}{10^6}$ meter = 10^{-6} meter = 1 micron.
1 nanometer = $\frac{1}{10^9}$ meter = 10^{-9} meter = 1 billionth meter.

Reconnaissance and ENERGY VISUALIZATION.

by Stephen Weber

Part III. Extensions to Our Vision.

Telescopes and Astronomy.

Chapter 26. Yerkes Observatory.

How did the largest refractor telescope in the world arrive in Williams Bay in 1892? Mark Twain called this period the Gilded Age. By an Act of Congress, Chicago was awarded the Columbian Exposition for 1893 to celebrate the 400th anniversary of Columbus arriving in America. The University of Chicago was founded in 1890 as the Harvard of the Midwest. Harvard University had a telescope. The University of Chicago needed a telescope. Astronomer George Ellery Hale convinced Chicago businessman Charles T. Yerkes to finance the observatory named in his honor for $400,000.00 to construct the telescope and build the observatory from1892 to 1897. While the observatory was being constructed on 77 acres overlooking Geneva Lake, the great refractor telescope was exhibited at the Columbian Exposition in Chicago. From 1897 to 1908 Yerkes Observatory in Williams Bay Wisconsin was "the birthplace of modern astrophysics" and the largest astronomical telescope in the world. Energy could be visualized light years away.

In 1903 George Ellery Hale left Yerkes to work on the Snow Solar telescope on Mt Wilson. In 1905 the Harvard 60 inch reflector telescope, built at the Harvard College Observatory became the largest telescope in the world. In 1908 the Hale 60 inch reflector telescope was completed at Mt. Wilson Observatory in Pasadena, California and was equal to Harvard's 60. The difference between the refractor and reflector telescope is that the refractor (Yerkes) uses 2 lenses, while the reflector (Harvard and Mt. Wilson) uses a lens and a metallic coated reflecting glass mirror.

Yerkes Observatory has a 40 inch refractor telescope astronomical observatory in Williams Bay, Wisconsin founded in 1892. It was owned by the University of Chicago Dept. of Astronomy and Astrophysics. The Yerkes Observatory was the largest refractor telescope using lenses to concentrate and direct light in the world at that time. The Observatory had labs for physics and chemistry and was the center for serious astronomical work for more than 100 years. The telescopes consisted of the 40 inch (102 cm) refractor dedicated in 1897, the 40 inch (102 cm) reflector, the 24 inch Cassegrain reflector, the 10 inch Cassegrain reflector, the 7 inch Schmidt camera, the 12 inch Kenwood Reflector, and the 23.5 inch refractor. The 77 acre campus also included a golf course which I loved to play, because you could see the lake from several holes. In addition to being a great observatory, Yerkes was a complete physical science center with physics and chemistry laboratories, machine shops, photographic workshops and laboratories.

The Yerkes great refractor is housed inside a clear span tower that is 112 feet high 90 feet in diameter. The telescopic tube is 64 feet in length and has two 40 inch diameter lenses, a convex lens – crown glass, and the other concave – flint glass, at the proper distance to bring light to a perfect focus 63.51 feet at the back end for perfect viewing and photographs. Yerkes has a great collection of glass photographic plates. The lenses weigh about 500 pounds and were cast in France in the late 1880's and were polished by the optical firm of Alvin Clark and Sons in Cambridge, Massachusetts. The telescope structure is constructed entirely of steel and was built by Warner & Swasey. The telescope stands 65 feet high from the base to the top of the German equatorial mounting, and rests on a concrete and brick pedestal about 25 feet high. The weight of the entire telescope is 82 tons and can be positioned by one person. The telescope is properly accessed by a moveable floor 75 feet in diameter and weighing 37 ½ tons that raises and lowers through a total distance of 26 feet using a counter weight system. It is the largest indoor elevator in the world. The Yerkes great refractor has been visited and used by some of the world's great scientists and astronomers as shown in the Yerkes photograph. There is a spiritual feeling of being closer to God when viewing the heavenly bodies light years away at Yerkes Observatory.

Figure 26.1 Einstein visited Yerkes in 1921 when this picture was taken, courtesy of Yerkes.

Yerkes Observatory has been home to Nobel laureates Albert Einstein, Edwin Hubble and Carl Sagen and important scientific achievements. After 123 years, Yerkes has changed its emphasis from original research to education. The University of Chicago sold Yerkes Observatory and 50 acres in September 2020 to Yerkes Future Foundation, PO Box 346,

Williams Bay, WI 53191. The Yerkes Future Foundation is a 501C3 Non Profit Corporation with EIN 83-0802129. The registered agent is Diana Colman. The new executive director is Dennis Kois with a Bachelors Degree from UW Milwaukee in Museum Design and a Masters Degree in Museum Studies.

Named for Charles Yerkes, a Chicago businessman who is credited with establishing the city's elevated train system and who subsidized the observatory, the facility is considered the birthplace of modern astrophysics. Yerkes is an astronomical university with 7 telescopes and labs for physical science. Its most prominent feature is the world's largest refractor telescope with a 40 inch lens and 63 foot tube.

In addition to the pair of Nobel laureates who worked at Yerkes, a third laureate was affiliated with the observatory. Albert Einstein visited , and Carl Sagan and Edwin Hubble studied there. Acclaimed landscape architect Frederick Law Olmsted, whose work includes New York City's Central Park, designed the grounds. But, over the last several decades, cutting-edge research has moved beyond Yerkes' equipment. The University of Chicago has invested in an Extremely Large Telescope (ELT) with a 39 meter diameter, sophisticated reflecting telescope project in Chile. Yerkes has remained a unique place that brought together formally trained scientists, astronomers, physicists, educators, students and more casual enthusiasts of astronomy and astrophysics. Yerkes future is in science and computer education to create expertise in space.

Figure 26.2 Gigantic Yerkes Observatory with brick masons scaffolds (left) for remodeling.

Author Steve Weber stands in front of Yerkes Observatory which is fenced off for the brick and masonry tuck point remodeling which started after Yerkes was closed to the public in 2018 to facilitate maintenance. The multi story scaffold for the brick mason contractors is on the left of this photo. The building is in fairly good condition considering that it is 123 years old and is exposed to the winter weather. The George Williams College of Aurora University is adjacent to Yerkes Observatory 77 acre campus. The Yerkes Future Foundation is consulting with staff at Lowell Observatory in Flagstaff, Arizona, which was established in 1894 and draws nearly 80,000 visitors a year, to find a marquee experience to attract enough visitors to generate significant funding. Perhaps that marquee experience is the 124 year history and the noble prize winners that visited and studied at Yerkes including: Einstein, Edwin Hubble and Carl Sagan.

The 40-inch refractor telescope towers over people during a tour of Yerkes Observatory on June 30, 2018, in Williams Bay, Wis. The University of Chicago, which owns the observatory and surrounding 77 acres of land along Geneva Lake, has announced it plans to sell the facility, which was built from 1893 to 1897. Funding is likely the highest hurdle. One foundation member said the organization would need $20 million to purchase Yerkes and maintenance could cost $400,000 to $800,000 a year. An accurate business funding plan is needed.

Figure 26.3 The Great 40 inch Refractor Telescope inside the Yerkes Observatory Dome.

None of the donors the foundation has approached has refused them outright. Are they hesitant because they want to see more significant plans? The Yerkes Future Foundation hired a museum director to turn Yerkes into a museum. The Art

Institute of Chicago has a world class museum upstairs and a world famous school downstairs. Yerkes has physics and chemistry labs and five additional major telescopes for students and instructors in addition to the main 40 inch refractor. Yerkes can be a school for the new Space Force and the colleges within a 75 mile radius as well as elementary and high school field trips in addition to being a historic astronomy museum.

Although the gates will be locked for maintenance, pedestrians and bicyclists will still have access to the property, university officials said. The school will continue maintaining the grounds and building. All five of Yerkes' telescopes will continue to operate, and the observatory will be open for research, the university stated on its website.

Author Stephen Weber stands in awe on the steps of the famous 123 year old Yerkes Observatory in Williams Bay, Wisconsin during the remodeling and repair phase. The author has a summer cottage across Lake Geneva from Williams Bay in the Wooddale subdivision and enjoyed playing golf at the beautiful golf course on the Yerkes Observatory grounds 20 years ago. The history of Yerkes is the great history of modern physics. So much has happened: nuclear physics, quantum mechanics, solid state physics, the atomic bomb, computers, lasers, aircraft, satellites, space shuttle, moon landings, the Hubble orbiting telescope, television etc.

Figure 26.4 Author Steve Weber at Yerkes front entrance with 40 inch telescope in background.

STEM (Science, Technology, Engineering, and Math) Education is the Key to Saving Yerkes. Author Steve Weber stands at the entrance to the historic 1893 Yerkes Observatory and says that education is the key to saving the famous Yerkes Observatory which was visited by three famous Noble Prize winners: Albert Einstein, Edwin Hubble and Carl Sagan. Georges Lemaitre, a Belgian catholic priest, mathematician and physics professor proved Einstein's general theory of relatively and proved mathematically that the universe is expanding which later became Hubble's Law.

Lemaitre provided an estimate of Hubble's constant. Edwin Hubble and Georges Lemaitre verified that the universe was expanding with the red shift phenomenon when stars are moving away. Hubble proved Lemaitre's theory with his analysis of the Andromeda Galaxy that was thought to be nebulae and was beyond the Milky Way galaxy. Lemaitre's theory became known as the Big Bang Theory. Lemaitre was nominated for the Noble prize in 1954 and 1956.

This is the marquee experience story that should be told to all visitors to Yerkes Observatory to fund the Yerkes Observatory in the future. This is a course in modern science history. This is a course in modern physics. This is a course in aviation and space history. This is a course in astrophysics and space exploration. This is the program to introduce students to study science. There could be science stations scattered around the 40 inch refractor telescope of major science developments and how they relate in time to Yerkes. The new US Space Force Program could explain the relevance of the Yerkes Observatory experience to future teachers and students.

Figure 26.5 Yerkes Observatory Front Entrance Cornerstones Dated 1893.

Reconnaissance and ENERGY VISUALIZATION.

by Stephen Weber

Part III. Extensions to Our Vision.

Telescopes and Astronomy.

Chapter 27. Hubble and Our Universe.

The Hubble Space Telescope was launched into low earth orbit in 1990 and still remains in operation 30 years later. The Hubble features a 2.4 meter (7.9 ft.) mirror and its 4 main instruments observe in the ultraviolet, visible and near infrared region of the electromagnetic spectrum. The Hubble telescope was named after astronomer Edwin Hubble and was built by the United States space agency NASA with contributions from the European Space Agency. The Goddard Space Flight Center controls the spacecraft. Hubble's orbit is outside of the distortion of earth's atmosphere which allows it to receive extremely high resolution images with lower background light than ground based telescopes. Hubble was launched into orbit by the Space Shuttle Discovery and is so large that it is designed to be maintained in space.

The advantages of a space base telescope is the angular resolution (the smallest separation at which objects can be clearly distinguished) would be limited only by diffraction, rather than by the turbulence in the atmosphere which causes stars to twinkle. In addition, a space based telescope could observe infrared and ultraviolet light which are hygroscopic and are absorbed by the water vapor in our atmosphere. There have been other space base telescopes. In 1962 NASA launched the Orbiting Solar Observatory (OSO) to also obtain UV, X-ray, and gamma-ray spectra. The first Orbiting Astronomical Observatory (OAO) was launched in 1962. In 1978 Congress approved $38 million funding for a space based telescope and the design began in earnest with a launch date of 1983 and the space telescope was named after Edwin Hubble. Perkin Elmer and Lockheed were behind schedule and over budget. The space shuttle that was to carry the space telescope was 30% over budget and three months behind schedule by 1985. In January 1986 the Challenger disaster brought the space shuttle program to a halt and grounded the shuttle fleet. The launch of the Hubble space telescope was delayed for several years. Hubble was finally launched in 1990 by space shuttle discovery and several weeks later flawed images showed that the primary mirror was polished in the wrong direction introducing spherical aberration and requiring compensation by using sophisticated image processing techniques until a new properly polished mirror could be installed with a space shuttle mission in 1993. Hubble is the only telescope that has to be maintained in space by astronauts. There have been five space shuttle missions to repair, upgrade, and replace systems including all 5 of the main instruments.

The Hubble Space Telescope (HST) carries five main scientific instruments:
1. The Wide Field/ Planetary Camera.
2. The Goddard High Resolution Spectrograph.
3. The High Speed Photometer.
4. The Faint Object Camera.
5. The Faint Object Spectrograph.

The Hubble Space Telescope (HST) and the scientific instruments are computer controlled and have been updated in 1993 to include 2 redundant strings of Intel based 80386 micro cpu's and an Intel 80387 math co-processor. In 1999 this was replaced with an Intel 80486 based system that was 20 times faster and with 6 times more memory. Some of the science instruments had their own embedded RCA 1802 microprocessor based systems. The wide field / planataray camera was built by NASA's Jet Propulsion Laboratory had 8 CCD arrays, 4 for each camera.

The Goddard High Resolution Spectrograph was designed to operate in the ultraviolet region and was built by the Goddard Space Flight Center and had 90,000 lines high resolution. The FOC and FOS were optimized for ultraviolet (UV) and used photon counting digicons as their detectors. The European space agency constructed the FOC, while the FOS was built by the University of California at San Diego and Martin Marietta.

The high speed photometer (HSP) was designed and built by the University of Wisconsin at Madison. It was optimized for UV and visible light observations of variable stars and objects varying in brightness and can take 100,000 measurements per second. The HST guidance system can also be used as a scientific instrument (x,y,z) to keep the telescope accurately pointed during an observation for extremely accurate astronomy to minimize roll, drift and pitch.

The Hubble Space Telescope is a satellite that requires ground support and is operated by AURA the Association of Universities for Research in Astronomy and is located in Baltimore, Maryland on the campus of Johns Hopkins University and the European Space Astronomy Center. There are many complex tasks such as scheduling observations for the telescope in a low earth orbit to facilitate service missions and avoiding radiation and exposure from the sun and keep bright light and scattered light from Hubble's 5 main astronomical instruments. The low earth orbit implies that target stars may be blocked from view by the earth for one half of each orbit. A successor to Hubble is the James Webb Space Telescope (JWST) scheduled for late 2021 launch. Hubble is still viable for 10 to 20 more years.

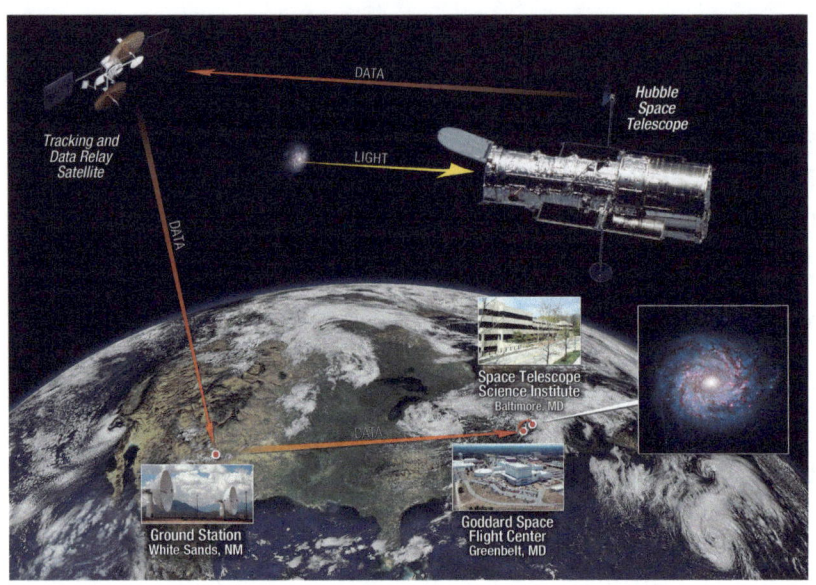

Figure 27.1 Hubble Space Telescope data transmission (NASA).

Figure 27.2 Hubble Space Telescope in orbit above earth. NASA

Figure 27.3 Hubble Space Telescope in operation above earth.

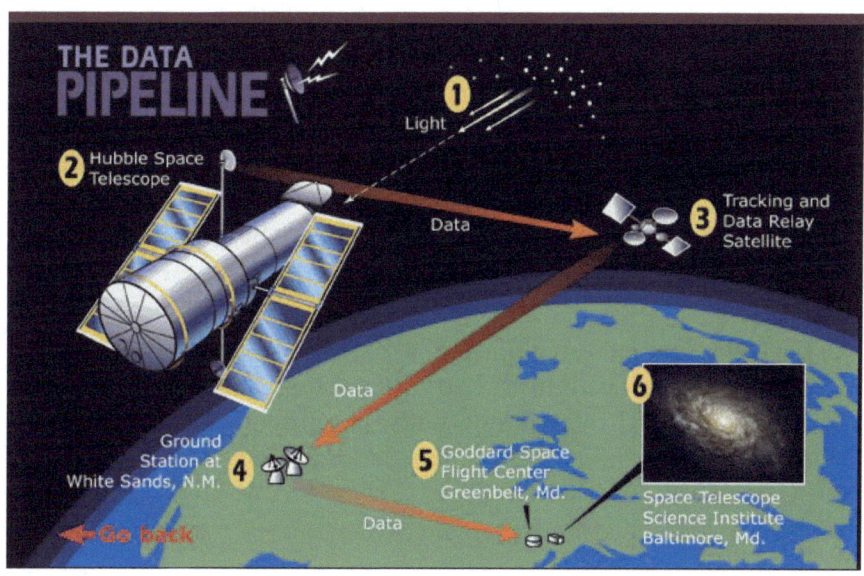

Figure 27.4 Hubble Space Telescope Data Pipeline orbiting above earth. NASA

Reconnaissance and ENERGY VISUALIZATION.

by Stephen Weber

Part III. Extensions to Our Vision.

Microscopes.

Chapter 28. Compound Microscope

Several sources say the microscope was invented in Holland around 1590 by Hans Liiershey.
There are several European inventors of the compound microscope in the Netherlands around 1620 by father and son lens makers Hans Martens and Zacharias Janssen.

Galileo Galilel was also credited with the invention of the microscope and presented his microscope invention to the Accedemia dei Lincei in 1625 and called it the "little eye".

Naturalists in Italy, England and the Netherlands began using microscopes to study biology. Leeuwenhoek achieved up to 300 times magnification using a simple single lens microscope. He could view red blood cells and spermatozoa and the discovery of micro-organisms.

In 1893 Kohler developed a sample method of illumination using an electric light to better see the specimen on the slide. The compound microscope with magnification range from 10 to 400 is the type used in school laboratories.

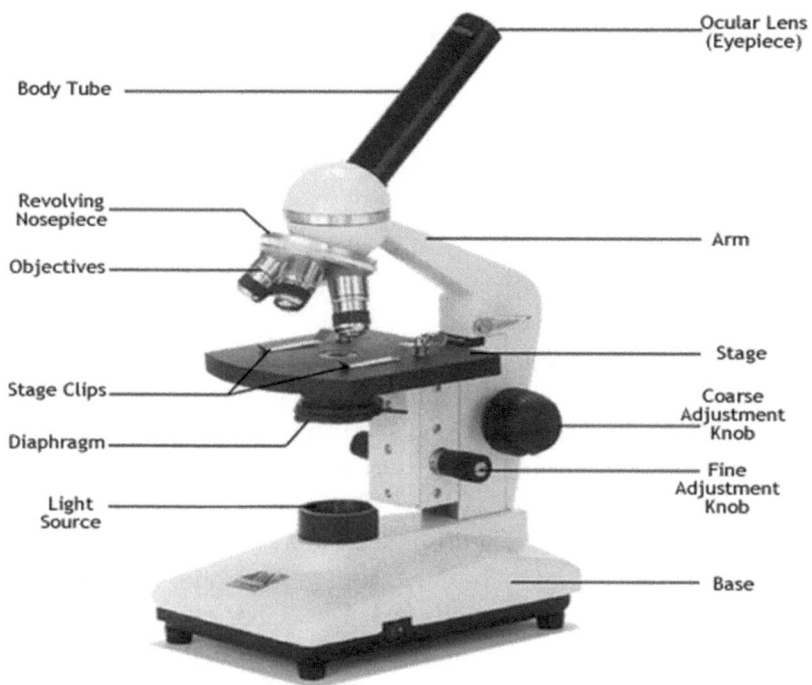

Figure 28.1 Compound Microscope with a 3 lens turret and a light source suitable for school labs.

Reconnaissance and ENERGY VISUALIZATION.

Part III. Extensions to Our Vision.

by Stephen Weber

Microscopes.

Chapter 29. Electron Microscope.

The electron microscope uses a beam of electrons to generate an image rather than light and electromagnets in the place of glass lenses. The transmission electron microscope allows for much higher resolution. This was developed in Germany in the early 1930's. After WWII they became commercially available via Siemens for discoveries and virus and harmful cell study, and pathogen detection. The scanning electron microscope was developed by Sir Charles Oatley and marketed by the Cambridge Electric Company. Then the scanning probe microscope was developed in the early 1980's by IBM in Zurich, Switzerland with quantum tunneling theory and later at AT&T at Murray Hill New Jersey and the Nobel Prize in Physics in 1986.

The fluorescence microscopes were developed for biology needs in targeted chemical staining to label and identify DNA, antibodies and in vaccine development to combat harmful virus. In 1978 Thomas and Christopher Cremer developed a practical laser scanning microscope. This was followed by the development of super resolution optical microscopes for analysis of fluorescently labeled samples and techniques for stimulated emission depletion could even approach the resolution of the electron microscope. The Nobel Prize in Chemistry for STED was awarded in 2014 for the development of the STED technique for fluorescence microscopy for single molecule visualization. This was followed by x-ray microscope development.

Reconnaissance and ENERGY VISUALIZATION.

by Stephen Weber

Part III. Extensions to Our Vision.

Photography and Television.

Chapter 30. The Camera and Photography.

There are two basic types of cameras: film and digital. Film cameras have been around since the civil war. The original cameras had a pinhole lens and a large plate of silver acetate for film. There were several types of cameras: the speed graphic for the press, the 2 ¼ Hasselblad and Rolliflex , and the 35mm cameras like Nikon with focal plane or regular shutters. These cameras are film cameras initially black and white and then mostly color either prints or slides. These great cameras had good lenses that could be focused, an iris to adjust the exposure and depth of field, and a shutter or focal plane shutter to control the duration of the exposure.

Most new cameras are digital. The digital camera revolution has largely replaced film cameras because you don't have to buy film. The digital camera has a CCD (charge coupled device) array to replace the film. If the CCD array measured 1000 pixels by 1000 pixels, this would measure one million pixels or one megapixel. Most digital cameras have more than this and 2 to 12 megapixel digital cameras are common. For example, if the array is 2.5 inches X 5 inches and the resolution was 1000 pixels per inch, that would be 2,500 pixels X 5,000 pixels = 12.5 megapixel camera.

Both the CCD and the newer CMOS chips convert light energy into electron flow and use a shift register technique to read it off one corner of the chip. CMOS technology uses lower voltage from 3.5 volts to 5 volts, whereas the CCD array uses 7 to 10 volts. CMOS technology is faster than CCD technology and can use parallel processing shift registers. CCD technology has greater noise immunity and is more sensitive to low light level. Because CMOS chips are cheaper to manufacture and use lower voltage, the newer digital cameras have come down in price and use the newer CMOS technology.

Reconnaissance and ENERGY VISUALIZATION

by Stephen Weber

Part III. Extensions to Our Vision.

Photography and Television.

Chapter 31. Motion Pictures.

In 1888 Thomas Edison began experimenting with a motion picture camera that when the pictures are played back at 24 frames per second they look like they are moving. At the same time in 1988 in England Edward Muybridge, a British photographer, tried to prove that when a horse ran all 4 of the horse's legs were off the ground. Muybridge took several photos very fast and played them back to show that the horse's hoof's were off the ground when running fast.

In 1889 Edison built a Strip Kinetoscope, which was an early movie camera and used a long flexible strip of flexible film that could be wrapped around a spool and when played back they seemed to move. The Edison team built a Kinetoscope and opened a movie theater in 1894.

In 1896 Thomas Armat and Francis Jenkins designed a phantascope which was an early movie projector that showed films on a screen so that a room full of people could view them. Edison bought the rights and started making his own projectors. The Lumiere brothers in France were also important in the development of movies. Edison is called the "Father of Motion Pictures."

The motion picture industry is a multi-billion dollar business. It started with black and white silent films and then in the 1920's they added sound and called them talkies. Walt Disney added music and sound to his Mickey Mouse cartoons and later synchronized the music to the video. Novels, music and movies exist in the time domain. Most movies are based on a story from a novel or a play and have a plot with a climax, a setting and a cast of characters.

The annual Oscar ceremonies celebrate the many famous movie stars, the films, the music, the directors, the supporting actors, the costumes, the make-up artists, the writers, the scene design etc. Of course the important issue is which film wins the Oscar for that year.

The technology has gone from 16mm movie cameras and projectors to 35mm cameras and projectors and then to 70mm movie cameras and projectors. Then the technology changed to video cameras and TV screens because it is much faster and cheaper. The film would have to be developed and edited, whereas, the video is right there to be edited. The video technology is there from the television industry.

Reconnaissance and ENERGY VISUALIZATION.

by Stephen Weber

Part III. Extensions to Our Vision.

Photography and Television.

Chapter 32. Television.

In 1927 Philo T. Farnsworth patented his invention of television. At the age of 14 after plowing a field on his parents' farm in Idaho, Farnsworth could visualize a picture made up of the rows of his neatly plowed field. Farnsworth envisioned painting the image a line at a time just like a line of type, line after line to compose a page or a picture. At that time Farnsworth was trying to figure out how to transmit a picture; and he came up with the idea of transmitting a line at a time to compose a picture at the age of 14.

Farnsworth had read many electrical and scientific magazines and he drew a picture of his image dissector tube. He presented the idea to his high school chemistry teacher along with a drawing of an electrical schematic, which he kept and later it was instrumental in winning a patent law suit in 1935 against RCA. The telegraph and radio were popular to transmit words and music. Marconi had invented the vacuum tube. Farnsworth used vacuum tubes in his designs for television to transmit a picture.

In 1923 Farnsworth moved to Provo Utah and studied mathematics and physics at Brigham Young University. Farnsworth worked for a community campaign and received funding for his electronic television idea. He moved and set up his lab in at 202 Green Street in San Francisco. With the help of his brother in law who did the glass blowing they perfected the image dissector tube which functioned like a TV camera. In September Farnsworth transmitted the first electronic television image and applied for his first television patent in 1927. Farnsworth was the first to form and manipulate an electron beam without mechanical devices to transmit images that contributed to breakthroughs in radar in WWII. At the beginning of the great depression his investors were looking for financial success and requested a meeting. Farnsworth took a picture of a Dollar sign with his image dissector tube in one room and displayed it electronically on his closed circuit television in another room to his investors.

In 1930 Farnsworth invited a former Westinghouse employee named Zworykin to visit his lab and see his television with the hopes of getting funding from Westinghouse. Zworykin later replicated Farnsworth's image dissector tube and presented it to RCA. Farnsworth refused to sell this invention to RCA. Consequently, RCA sued Farnsworth for priority of invention to get the commercial television rights. In 1935 Farnsworth won the law suit.

The TV timeline starts in 1927 with Farnsworth's patent to the present about 93 years. After 17 years Farnsworth's patent expired and the FCC worked to establish broadcast standards. Both Farnsworth and RCA had over 200 patents each. By the 1950's there were 3 major broadcast stations: CBS, NBC and ABC. The broadcast tower antenna's for each station are separated by 6 MHz and transmit the aural or audio which is FM modulated separately from the video which is AM modulated and fed to an antenna via a diplexer. The signal was broadcast to a TV receiver.

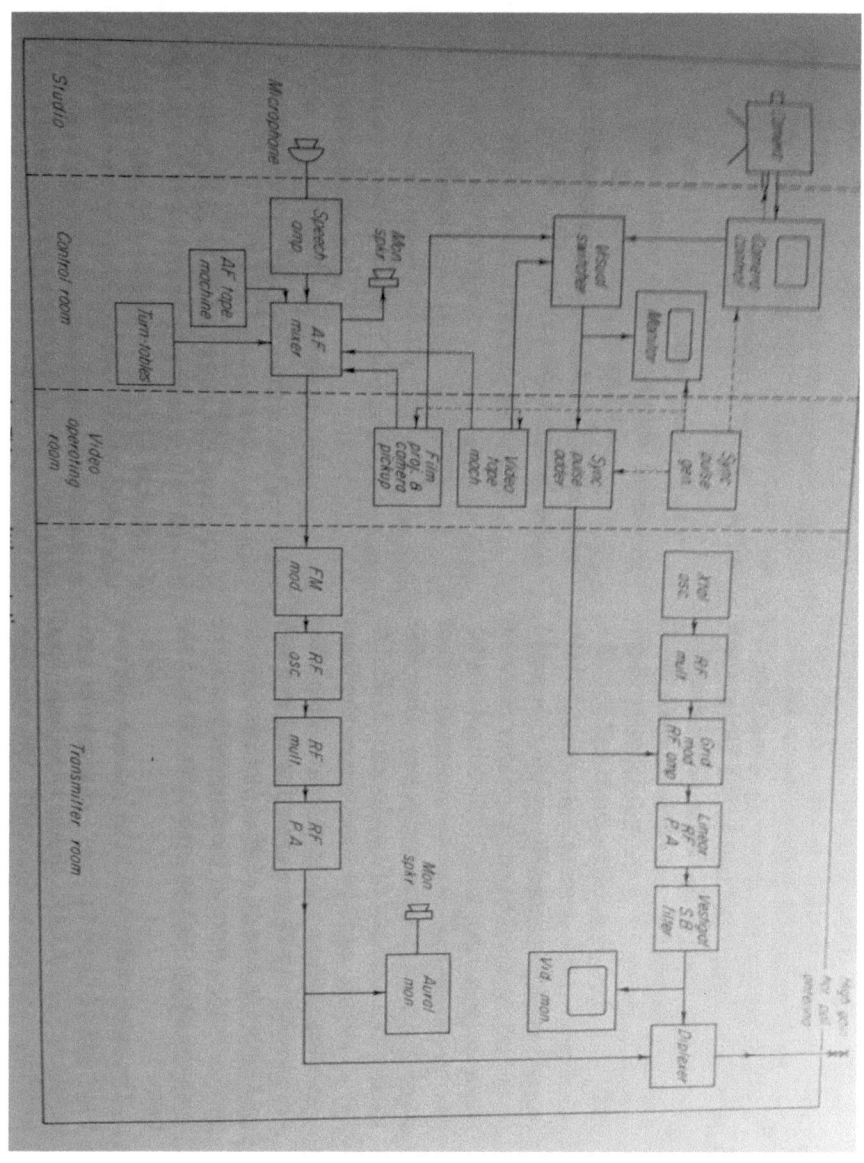

Figure 32.1 Television transmitting station block diagram.

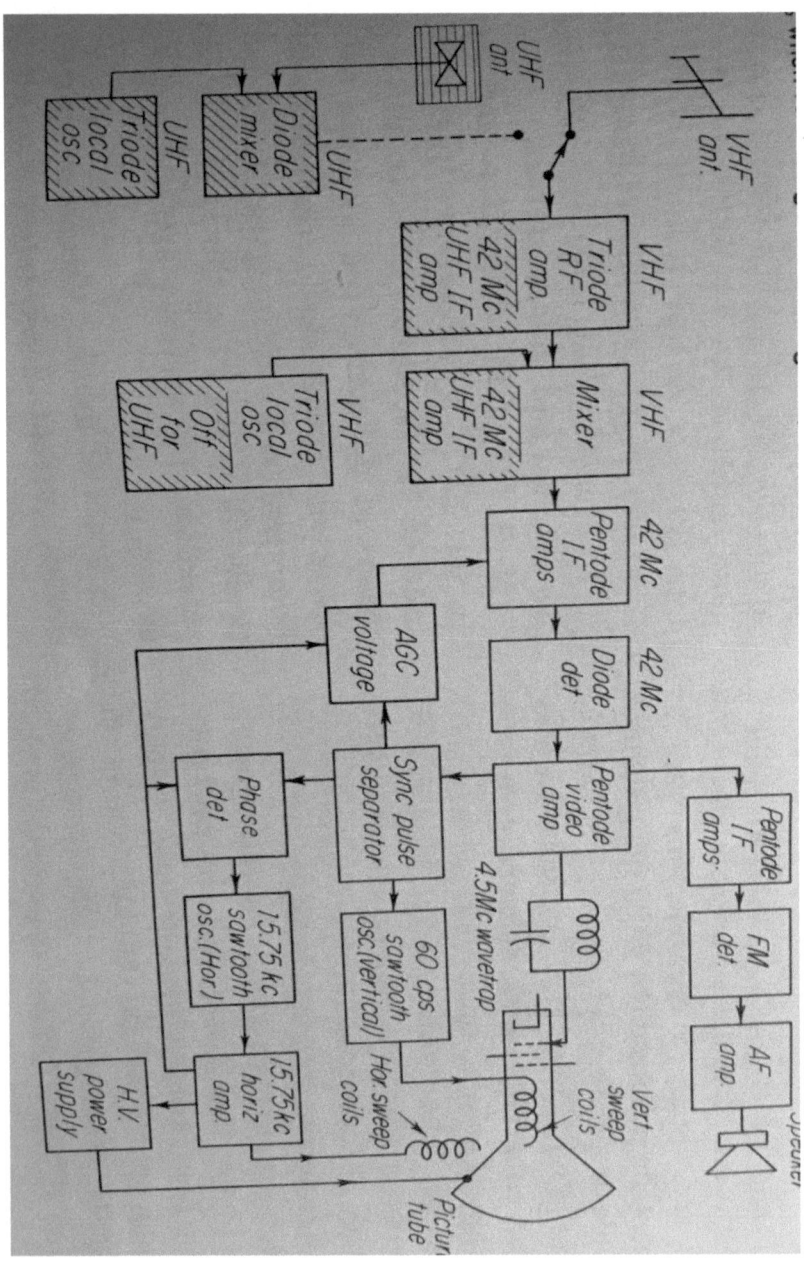

Figure 32.2 Television receiver block diagram.

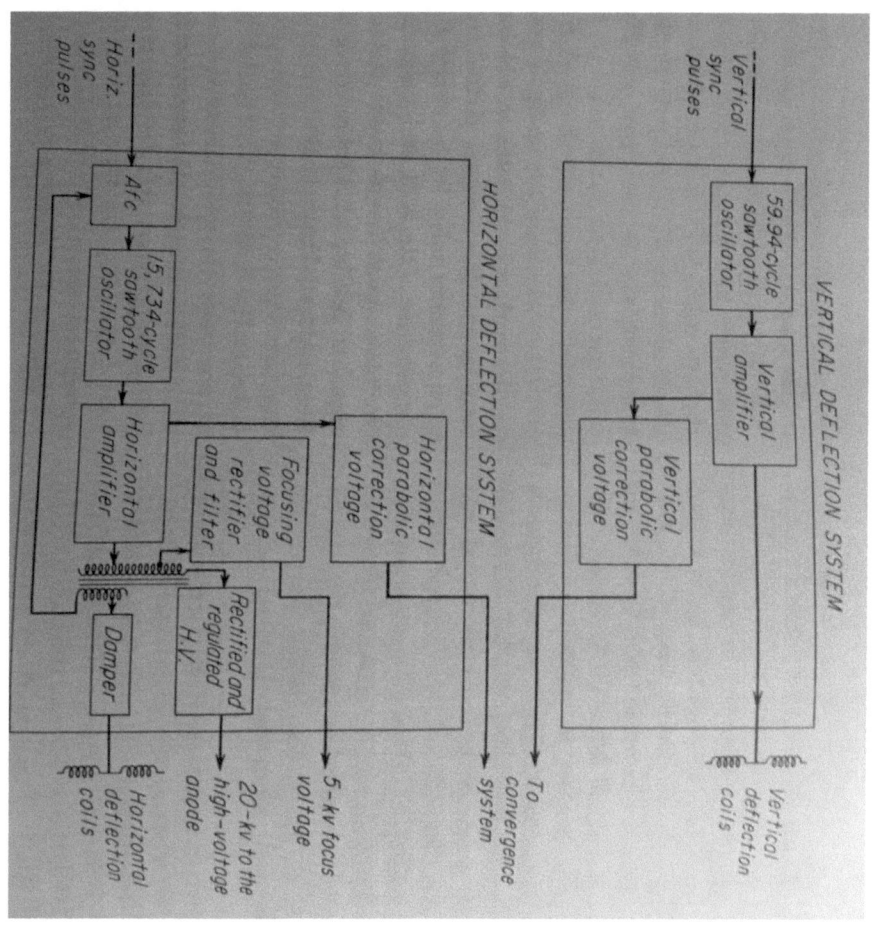

Figure 32.3 Television Horizontal and Vertical Deflection System.

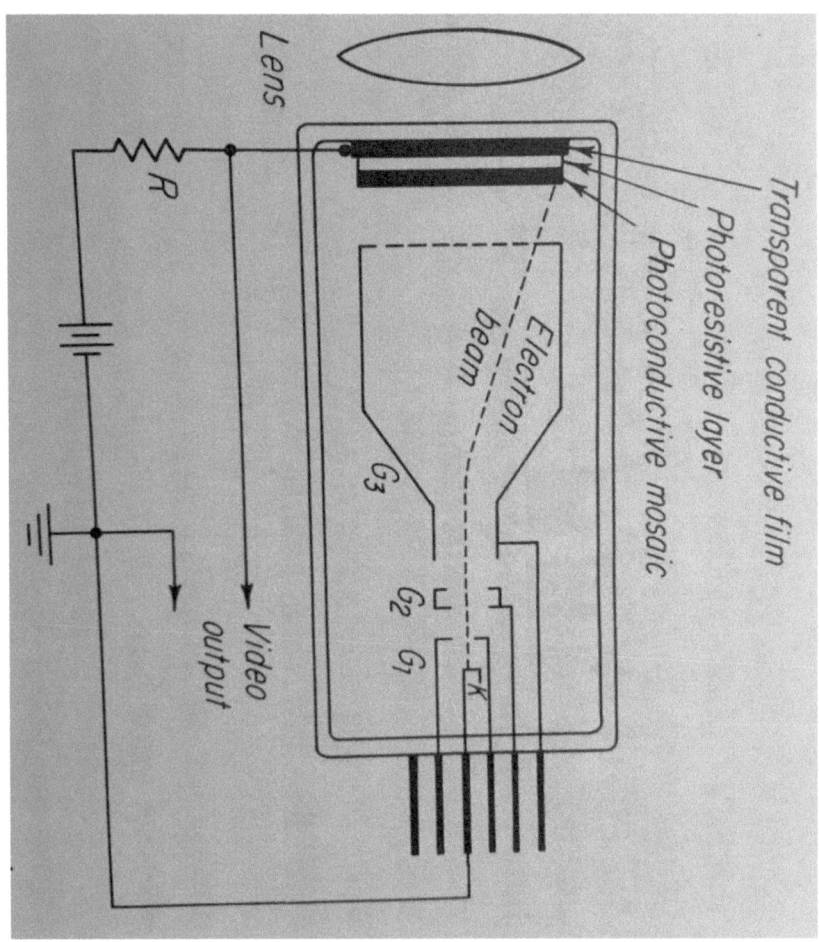

Figure 32.4 Television vidicon camera block diagram.

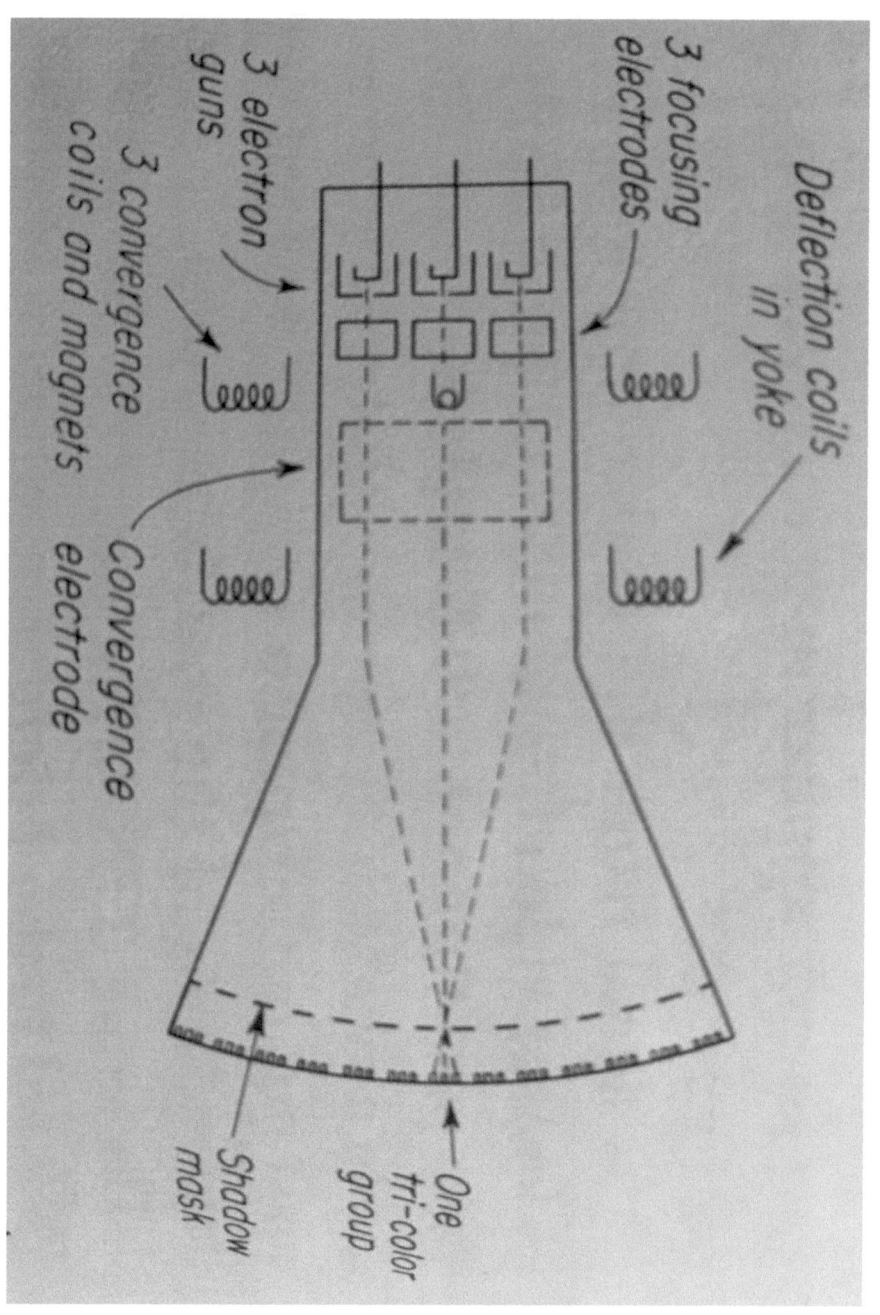

Figure 32.5 Three gun color television tube.

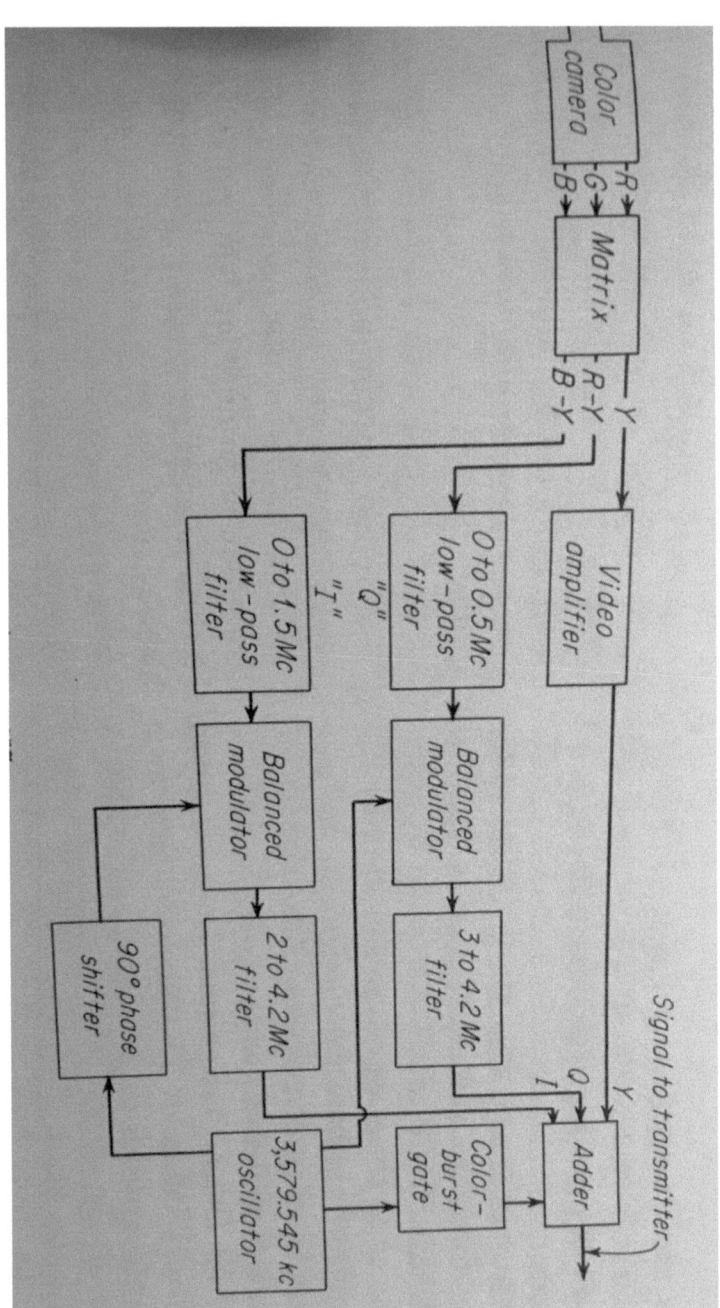

Figure 32.6 Color television circuit for RGB color transmission.

There were 3 television manufacturers in the Chicago area: Zenith, Motorola and Admiral. The glass picture tube for Zenith was made by Rauland in the 50's, 60's, 70's, 80's and 90's. After serving in the Air Force, I worked for Zenith in the late 1960's on military engineering projects in Don Robertson's Video Data Group, before going back to school on the GI Bill. I can remember going with Bill Leyseth, a Zenith project engineer, to Argonne and to the Rauland facility, which was so extensive, we drove around in a golf cart and watched the large TV tubes hanging from above and moving in a winding assembly line around Rauland.

Later on I worked as a project engineer at Profexray in medical x-ray on image intensifier tubes to minimize the amount of radiation to the patients. Although color television was first introduced in 1950, it was not until the 1960's that color television really took off. Even Japan was making color televisions with Sony and Panasonic.

By the year 2000 the technology was changing from analog to digital television with PCM, pulse code modulation transmission. Set top boxes were introduced to convert the digital signal to analog for display on the older analog televisions.

Then flat screen HDTV was introduced to properly display the digitally transmitted signal. Resolution increased from 525 lines to 720 lines to 1080 lines. The TV screen size increased from 20, 24, and 27 inch tube displays to 32, 40, 50, 60 and 70 inch diagonal flat screen displays. These were entertainment centers.

Reconnaissance and ENERGY VISUALIZATION.

by Stephen E. Weber

Part III. Medical Imaging Systems.

Chapter 33. Medical X-Ray Systems.

A Dutch physicist working in Germany at that time by the name of William Roentgen accidently discovered X-Rays, when electrons accelerated through a large potential difference collided with a metal target. The potential difference could be about 45,000 volts, and the metal target could be platinum or molybdenum as an example.

The target is contained within an evacuated glass tube, which was the custom for making radio vacuum tubes. The xy diagram of X-Ray intensity versus the wavelength is a broad continuous spectrum a Bremsstrahlung or braking radiation, with sharp peaks called characteristic X-ray lines, K_α and K_β because they involve the n=1 or atom K shell.

The spectral lines emitted when atoms give off electromagnetic radiation are due to transitions of atomic electrons between the initial and final stable electron states. The fall from energy level W_1 to W_2 results in the emission of a photon of with energy hv. $W_1 - W_2$ = hv, where h= Planck's constant.

The decreasing energy level spacing is implied in Bohr's formula for the individual permitted energy levels and de Broglie was the first to show that there is a class of photoelectrons ejected by X-rays of frequency f whose kinetic energy W_{kin} = hv-hv_q accurately satisfies the equation.
The Bohr-Heisenberg uncertainty principle implies that radiation and matter act like particles and have the characteristics of waves. The uncertainty principle states that

there is no known experiment that can be devised that can measure the momentum p and the position q of a system to a precision such that Δp Δq > h.

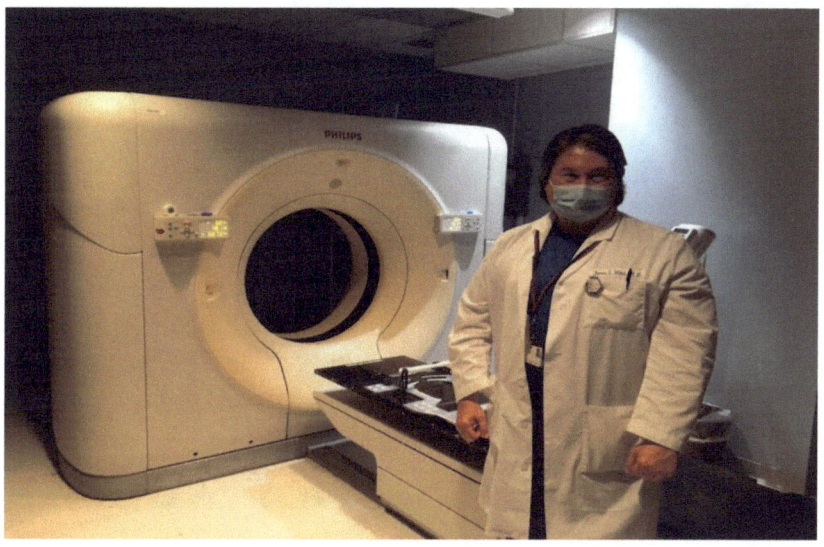

Figure 33.1 Dr. Walsh, Chairman of the Radiation Oncology Dept at Edward Hines Jr Hospital.

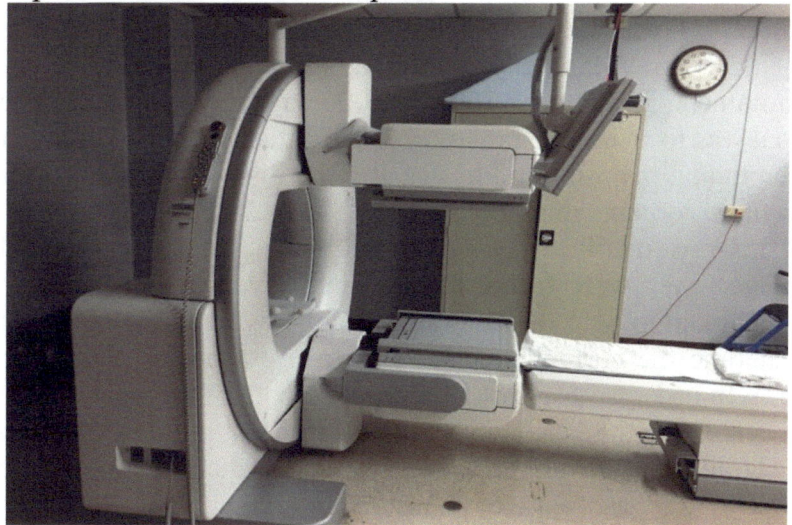

Figure 33.2 The nuclear medicine system used with technesium 99 for bone scans at Hines VA.

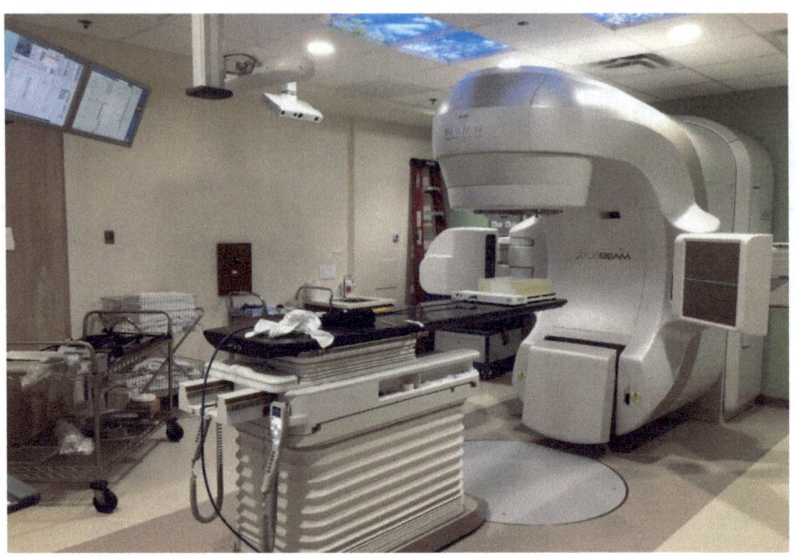

Figure 33.3 The new Varian Linear Accelerator in Radiation Oncology at Edward Hines Jr. Hospital.

In 1971 I was hired as a project engineer for Profexray, a medical X-ray company, that made a complete line of X-ray equipment, including a 100KV, 2000ma X-ray generator. The goal was to minimize radiation to the patient. After working at Zenith and majoring in Physics at NIU, I got a job as a project engineer on the imaging system. Medical imaging systems, like x-ray, require a new way of looking at an anatomical image. They were trying to develop a video imaging system for angiography to inject a radio opaque dye into the heart. At that time the current practice was to use a bucky plate with film. In order to display and record the path of radiopaque dye injected into the heart via a catheter inserted into the groin an image intensifier tube is used to minimize the radiation to the patient. I was sent to Yale New Haven hospital with Profexray's angiographic medical X-ray system that used a video camera interfaced to the image intensifier tube to minimize radiation to the patient and display everything on a TV monitor and record everything on video tape for subsequent study and review.

Rontgen discovered X-rays in 1895 accidentally. Later on, this technique was being used for medical investigation purposes. Light waves or radio waves are used in this modality which can pass through lighter objects like air, but they cannot pass through hard and denser objects like bone. A film is placed behind the object, which image is required to obtain and X rays are directed on the object. X rays pass through the air which is lighter but they are not passed through the denser object,

The main difference between X-ray and CT scan is that x-ray is used to detect bone fractures or joint dislocations while a CT scan is an advanced technique used to detect delicate soft tissues and internal organ injuries. The principle of the X-ray machine is to use light or radio waves as radiation for the detection of fractures of bone, dislocation of joints and somehow soft tissue or organ anomalies like lung infections, pneumonia etc. CT scan is said to be an advanced X-ray machine which provides a much detailed view of internal body organs and tissue anomalies.

CT scans can expose you to as much radiation as 200 chest X-rays. CT emits a powerful dose of radiation, in some cases equivalent to about 200 chest X-rays according to Consumer Reports.

CT scans and MRIs are both used to capture images within your body. The biggest difference is that MRIs (magnetic resonance imaging) use radio waves and CT (computed tomography) scans use X-rays. PET scans (positron emission tomography scans) are often done in conjunction with CT scans (computerized tomography scans) or MRI scans (magnetic resonance imaging scans).

While CT and MRI scans show images of your body's internal organs and tissues, PET scans can give your healthcare provider a view of complex systemic diseases by showing problems at the cellular level.

Unlike MRIs, PET scans use positrons. A tracer is injected into your body that allows the radiologist to see the area scanned. An MRI scan can be used when your organ shape or blood vessels are in question, whereas PET scans can be used to view your body's function.

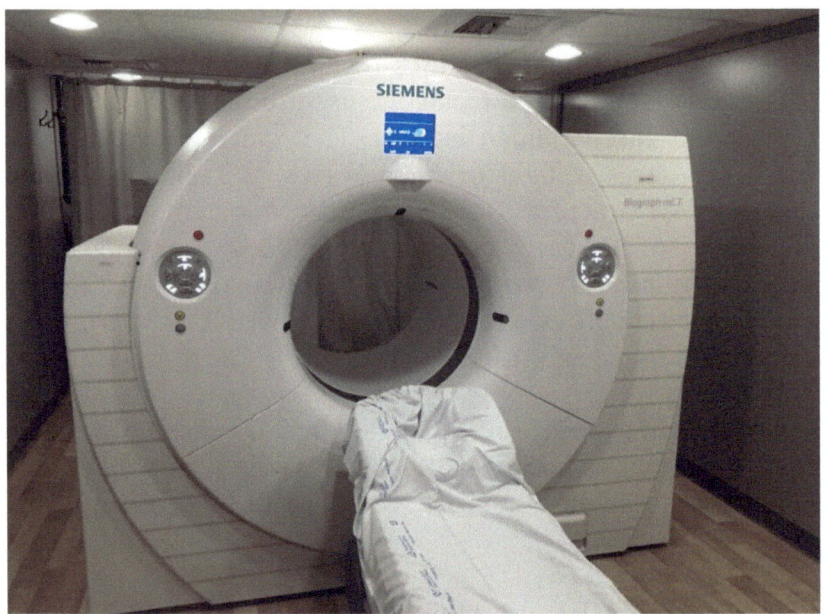

Figure 33.4 Siemens Biograph mCT PET Scan system at Edward Hines Jr. Hospital.

The PET Scan looks at function down to the cellular level. For example for a prostate specific PET Scan the patient is injected with an isotope such as F18 Sluciclovine which has an affinity for the prostate cancer.

After an IV injection the patient is placed in position and the radiation technologist has 5 minutes to do the whole exam the isotope travels through the body that fast. The PET scan picks up the radiation from the F18 sluciclovine isotope that lights up any cancer cells in the area affected. Thus the PET Scan records the areas illuminated by the isotope. The PET Scan is much higher resolution than the CT Scan and shows much more detail.

Reconnaissance and ENERGY VISUALIZATION.

by Stephen E. Weber

Part III. Medical Imaging Systems.

Chapter 34. Linear Accelerator, CT and PET Scan Systems

A computerized tomography (CT) or Computerized Axial Tomography (CAT) scan combines a series of X-ray images taken from different angles around the body and uses computer processing to create cross-sectional images (slices) of the bones, blood vessels and soft tissues inside the body. CT scan images provide more-detailed information than plain X-rays do. A CT scan has many uses, but it's particularly well-suited to quickly examine people who may have internal injuries from car accidents or other types of trauma. A CT scan can be used to visualize nearly all parts of the body and is used to diagnose disease or injury as well as to plan medical, surgical or radiation treatment.

During a CT scan, the patient is exposed to ionizing radiation. The amount of radiation is greater than one would get during a plain X-ray because the CT scan gathers more detailed information. The low doses of radiation used in CT scans have not been shown to cause long-term harm, although at much higher doses, there may be a small increase in the potential risk of cancer. CT scans have many benefits that outweigh any small potential risk. Doctors use the lowest dose of radiation possible to obtain the needed medical information. Also, newer, faster machines and techniques require less radiation than was previously used.

In certain cases, the doctor may recommend that the patient receive a special dye called contrast material. This can be something that is given to drink before the CT scan, or something that is given through a vein in the arm or inserted into the rectum. Although rare, the contrast material can cause medical problems or allergic reactions. Most reactions are mild and result in a rash or itchiness. CT scanners are shaped like a large doughnut standing on its side.

The patient lies on a narrow, motorized table that slides through the opening into a tunnel. Straps and pillows may be used to help the patient stay in position. During a head scan, the table may be fitted with a special cradle that holds the head still. While the table moves you into the scanner, detectors and the X-ray tube rotate around. Each rotation yields several images of thin slices of the body. A technologist in a separate room can see and hear everything. The patient will be able to communicate with the technologist via intercom. The technologist may ask the patient to hold his or her breath at certain points to avoid blurring the images.

After the exam the patient can return to a normal routine. If the patient was given contrast material, the patient may receive special instructions. In some cases, he may be asked to wait for a short time before leaving to ensure that he feels well after the exam. After the scan, he will likely be told to drink lots of fluids to help the kidneys remove the contrast material from the body. CT images are stored as electronic data files and are usually reviewed on a computer screen. A radiologist interprets these images and sends a report to your doctor. Computerized axial tomography (CAT) scan machines produce X-rays, a powerful form of electromagnetic energy.

X-ray photons are basically the same thing as visible light photons, but they have much more energy. This higher energy level allows X-ray beams to pass straight through most of the soft material in the human body. A conventional X-ray image is a shadow, a silhouette of the bones. In a CAT scan machine, the X-ray beam moves all around the patient, scanning from hundreds of different angles. The computer takes all this information and puts together a 3-D image of the body.

The CAT machine looks like a giant doughnut tipped on its side. The patient lies down on a platform, which slowly moves through the hole in the machine. The X-ray tube is mounted on a movable ring around the edges of the hole. The ring also supports an array of X-ray detectors directly opposite the X-ray tube. A motor turns the ring so that the X-ray tube and the X-ray detectors revolve around the body. Each full revolution scans a narrow, horizontal slice of the body. The control system moves the platform farther into the hole so the tube and detectors can scan the next slice.

In this way, the machine records X-ray slices across the body in a spiral motion. The computer varies the intensity of the X-rays in order to scan each type of tissue with the optimum power. After the patient passes through the machine, the computer combines all the information from each scan to form a detailed image of the body.

It's not usually necessary to scan the entire body, of course. More often, doctors will scan only a small section. Since they examine the body slice by slice, from all angles, CAT scans are much more comprehensive than conventional X-rays. Today, doctors use CAT scans to diagnose and treat a wide variety of ailments, including head trauma, cancer and osteoporosis. They are an invaluable tool in modern medicine.

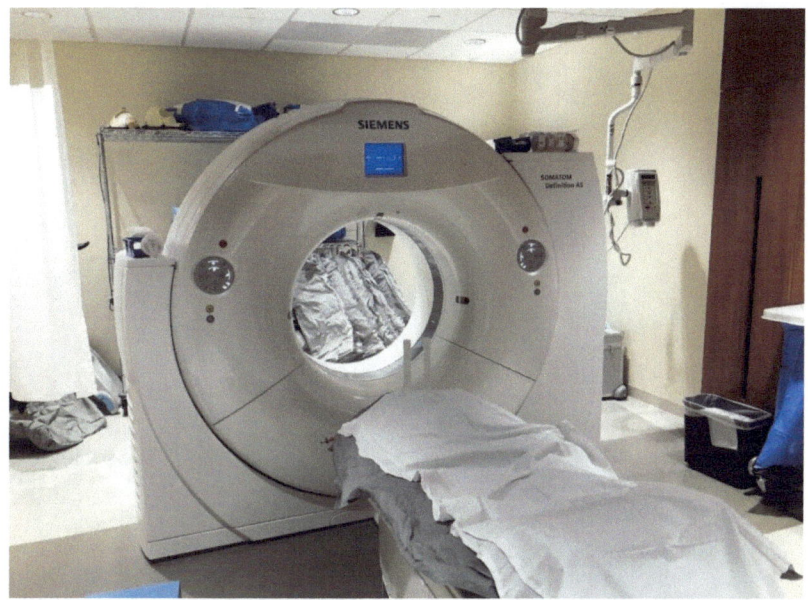

Figure 34.1 The Siemens CT Scan Simulation System at Loyola University Medical Center.

The Siemens CT Scan Simulation System at Loyola University Medical Center in Maywood is used to plan the patient radiation scenario to eradicate the cancer while safely radiating the patient. A mold is designed for each patient for precise alignment for successive treatments.

The patient is also given a tattoo. A Calypso system was previously used for exact alignment. Depending on the type of radiation therapy, three radiation seeds could be injected for triangulation. The radiation technicians fabricate the mold on the CT system so it can be moved to a Linear Accelerator for the prescribed radiation oncologist's repeated radiation treatments.

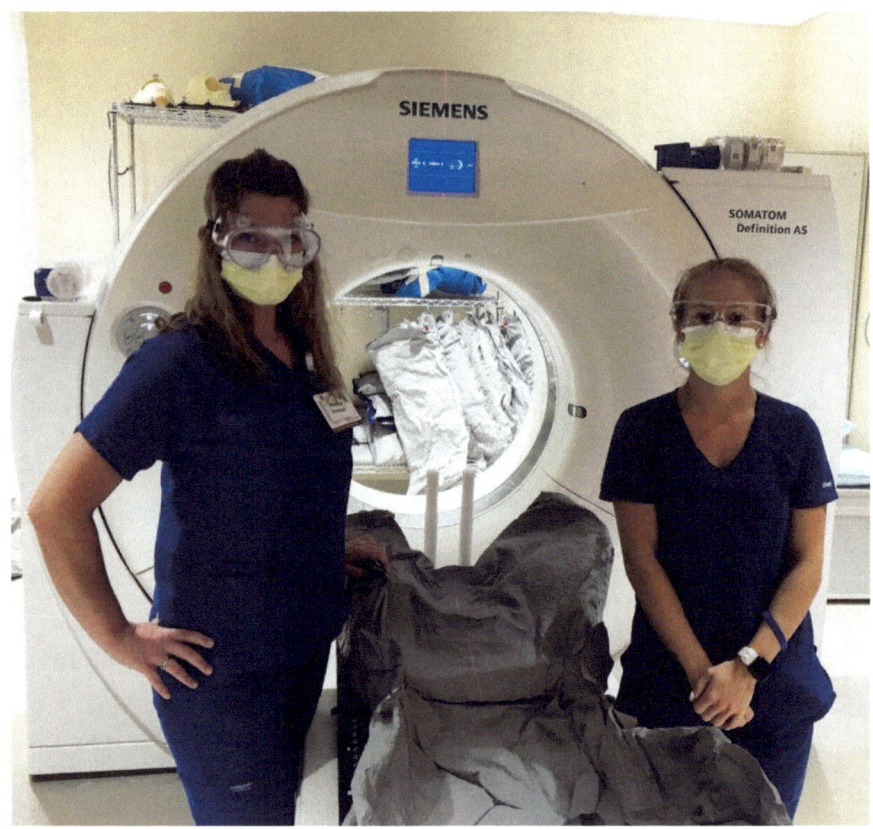

Figure 34.2 Siemens Somatom Definition AS at Loyola University Medical Center.

Radiation technicians Brandy and Jasmine are shown with the Siemens CT Scan System used for radiation simulation study by the radiation oncologist doctor to plan the angles and amounts of radiation necessary to kill the cancer while saving the patient's life. The mold is uncomfortable until it is inflated and magically hardened to the desired position for the patient.

The mold for that patient is brought to the linear accelerator for that patient's number of treatments. The radiation is cumulative to destroy the cancer and dosimeter badges are used by the radiation technologists for safety.

The Linear Accelerator at Loyola University is a very sophisticated system that can do a preview picture before radiation to insure proper alignment. Then when radiation starts the rotating arm can radiate while rotating to insure proper radiation to the patient depending on the plan.

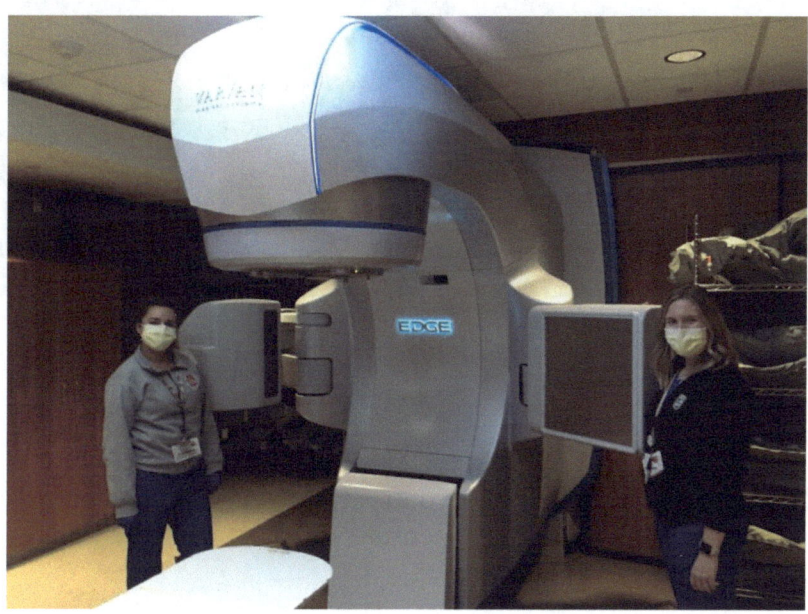

Figure 34.3 The Varian Edge Linear Accelerator at Loyola University Medical Center.

The molds for each patient are shown to the right on the shelves. The molds interlock to the radiation bed and fit precisely to each patient for accurate radiation.

Figure 34.4 A Loyola University Medical Center control room.

Each Loyola University Medical Center linear accelerator and CT scan system has its own control room with all the monitors to visualize progress and ensure quality control for the patient's radiation plan.

ENERGY VISUALIZATION II - Art & Science.
By Stephen Weber
Part I. Aerial Reconnaissance Acknowledgements:

Table 25.1 Computer Memory and Distance.
Table 25.2 Exponent Operations:
Figure 26.1 Einstein visited Yerkes in 1921 courtesy Yerkes.
Figure 26.2 Gigantic Yerkes Observatory with scaffolds (left)
Figure 26.3 The Great 40 inch Refractor Telescope Yerkes.
Figure 26.4 Author at Yerkes front entrance with telescope.
Figure 26.5 Yerkes Observatory Front Entrance Dated 1893.
Figure 27.1 Hubble Space Telescope transmission (NASA).
Figure 27.2 Hubble Space Telescope in orbit above earth.
Figure 27.3 Hubble Space Telescope in operation above earth.
Figure 27.4 Hubble Space Telescope Data Pipeline orbiting.
Figure 28.1 Compound Microscope with a 3 lens turret.
Figure 32.1 Television transmitting station block diagram.
Figure 32.2 Television receiver block diagram.
Figure 32.3 Television Horizontal and Vertical Deflection.
Figure 32.4 Television vidicon camera block diagram.
Figure 32.5 Three gun color television tube.
Figure 32.6 Color television circuit for RGB color transmit.
Figure 33.1 Dr. Walsh, Chairman of the Radiation Oncology
    Dept at Edward Hines Jr Hospital.
Figure 33.2 The nuclear medicine system used with
    technesium 99 for bone scans at Hines VA.
Figure 33.3 The new Varian Linear Accelerator in Radiation
    Oncology at Edward Hines Jr. Hospital.
Figure 33.4 Siemens Biograph mCT PET Scan system at
    Edward Hines Jr. Hospital.
Figure 34.1 The Siemens CT Scan Simulation System at
    Loyola University Medical Center.
Figure 34.2 Siemens Somatom Defination AS at Loyola
    University Medical Center.
Figure 34.3 The Varian Edge Linear Accelerator at Loyola
    University Medical Center.
Figure 34.4 A Loyola University Medical Center control room

Reconnaissance and ENERGY VISUALIZATION

By Stephen E. Weber

Part IV. Complex Systems and Problems.

Chapter 35. Solar Radiation.

The sun is the primary source of energy for the earth. The sun is 93 million miles from the earth. Because of this great distance the sun's rays reaching earth are considered parallel. The earth is tilted on an axis and orbits around the sun once per year or every 365 days. The sun is a nuclear fusion reactor. The energy of the sun is 3.8×10^{26} watts and radiates outward from the center of the sphere. The energy absorbed by the earth is 1.7×10^{11} watts. Solar energy traveling at the speed of light toward the earth in the vacuum of space is not attenuated until it reaches the earth atmosphere. The mass of the sun is 2×10^{30} kilograms (kg). The metric distance from the sun to the earth is 1.5×10^{11} meters. The temperature at the center of the sun is 40 million Kelvin. The surface temperature of the sun is 5,800 Kelvin. The solar radiation or incident energy from the sun is calculated in accordance with the Stefan-Boltzmann law $J = \alpha T^4$.

Temperature calculation: let Geq = the average equilibrium temperature of the earth,

then $\frac{dGeq}{dt} = \varepsilon \sigma T^4 A$, $\varepsilon = 1361 \frac{W}{m^2}$,

σ = Stefan Boltzmann 5.670×10^{-8} w/m^2 K, $A = \pi (r_e)^2$

$\frac{dGeq}{dt} = \varepsilon \sigma T^4 A = (1361 \frac{W}{m^2})(5.670 \times 10^{-8}$ w/m^2 K$)(T^4)(3.14(6,378,000))^2$

$$\frac{dGeq}{dt} = 1.74 \times 10^{17} \text{ Watt} \frac{\text{joule}}{\text{sec}}$$

$$T = \sqrt[4]{\left. 1.74 \times 10^{17} w \middle/ (0.95) x\ 5.6704 \times 10^{-8} \frac{w}{m^2} K\ (4\pi)(6{,}378{,}000)^2 \right.}$$

$$T = \sqrt[4]{\left. 1.74 \times 10^{17} w \middle/ (0.95) x\ 5.6704 \times 10^{-8} \frac{w}{m^2} K\ (12.567)4.54 \times 10^{13} \right.}$$

$$T = \sqrt[4]{\left. 1.74 \times 10^{12} w \middle/ (0.95) x\ 5.6704 K\ (12.567)4.54 \right.}$$

$$T = \sqrt[4]{\left. 1.74 \times 10^{10} \middle/ 3.071728262 \right.} = 274.34K - 273 = 1.34°C$$

Thus the earth's average calculated temperature is 1.34°C above the freezing point of water. A few degrees means a lot when global warming is considered. The measured temperature is colder, about 1.0 °C perhaps because the dust particles on top of the clouds reflect the sun's rays.

The emissivity e (0-1) can be tuned to 0.955 between reflected and absorbed towards measured.

$$T = \sqrt[4]{\left. 1.74 \times 10^{12} w \middle/ (0.955) x\ 5.6704 K\ (12.567)4.54 \right.}$$

$$T = \sqrt[4]{\left. 1.74 \times 10^{10} \middle/ 3.23339817 \right.} = 273.98K - 273 = 0.98°C$$

There are several techniques and formulas available for calculating solar radiation to the earth.
The estimate of the solar energy available at the earth's surface must take into consideration the solar constant I_{sc} = 1367 Watts / m^2
The earth's disc area is given by $4\pi R^2/2$ = 684 W/m^2.
Assuming that the sun shines 12 hours per day and 30% of the sun's energy is lost (atmosphere).
The unit Radiation = 0.7 x 684 W/m^2 x12 hours = 5.75 kW hours/ day.

There is an assumption that the amount of energy warming the earth is equal to the energy lost at night to keep the warmth of the earth the same. However, with global warming it is very difficult to calculate or measure the average annual temperature of the earth over a period of a year. The radius of the earth is 3,959 miles or 6,371km. The Energy received by the earth is given by
E_r = 1367 Watts / m^2 x π (Radius earth)^2 = 1367 x 3.1416 x $(6,371,000)^2$ = 173.5 x 10^{15} watts.

Some of the Energy received, E_r is reflected away and not absorbed. If the earth was covered with snow and ice, most of the energy would be reflected away. If the earth was dark, like a black body, most of the energy would be absorbed. The sunlight absorbed heats the earth. A black body emits electromagnetic radiation in the form of infrared radiation or IR in the infrared spectrum. The amount of radiation emitted by an object depends on the temperature of the object. The hotter, the greater the IR radiation! This relationship is the Stefan-Boltzmann law. J = αT^4 , where J= joules /sec /square meter or energy per unit time per unit area. α = Stefan Boltzmann constant = 5.670373 x 110^{-8} watts / m^2 K^4 , where K is the temperature in Kelvin. This Stefan Boltzmann law explains how much infrared energy the earth will emit per unit area.

As the earth rotates, all the surface is heated by sunlight. Thus the whole surface emits Infrared IR energy. The Energy emitted $E_e = \alpha T^4 \times 4\pi RE^2$. If IE is the fraction of incident energy reflected away from the earth, then (1- IE) is the amount of incident energy absorbed by the earth. Setting the energy in equal to the energy out or energy absorbed = energy emitted, from the conservation of energy law. Then: I_{SC} x (1- IE) x $\pi RE^2 = \alpha T^4 \times 4\pi RE^2$, where I_{SC} = the solar constant and α = the Stefan Boltzmann constant. Then: I_{SC} x (1- IE) = 4 αT^4. The earth's average IE or emissivity e is a variable from 0 to 1. Let us start with an estimate = 0.1, then we can solve for earth's temperature T.

$T = \sqrt[4]{I_{SC} \times (1 - IE)/4\alpha}$ solving for $T = \sqrt[4]{1367 \times (1 - 0.1)/4 \times 5.6704 \, x10^{-8}}$

$T = \sqrt[4]{1230.3/22.68 \, x10^{-8}} = \sqrt[4]{5424603175} = 271.38K - 273 = -1.611^0$ C

$T = \sqrt[4]{1367 \times (.92)/4 \times 5.6704 \, x10^{-8}} = \sqrt[4]{1257.64/22.68 \, x10^{-8}} = 272.89K - 273 = -0.11^0$ C

$T = \sqrt[4]{1367 \times (.93)/4 \times 5.6704 \, x10^{-8}} = \sqrt[4]{1271.31/22.68 \, x10^{-8}} = 273.62K - 273 = 0.62^0$ C

$$T = \sqrt[4]{1367 \times (.935) / 4 \times 5.6704 \, x10^{-8}} =$$
$$\sqrt[4]{1278.145 / 22.68 \, x10^{-8}} = 273.989 K - 273 = 0.98^0 \, C$$

Where k = kelvin, $T_{celsius} = T_{kelvin} - 273 = 0.98^0$ C,

The main idea of this chapter is that solar radiation from the sun is either reflected or absorbed and that the absorbed radiation warms the earth and this warmth can be calculated with the Stefan Boltzmann Law of Radiation $Q = \varepsilon \sigma T^4 At$. The radiant energy Q, emitted in a time t by an object that has a Kelvin temperature T, a surface area A and an emissivity e from (0 to 1). The emissivity depends on the surface, for example for a black body the emissivity would be 1. By adjusting the emissivity the calculated temperature is = to the measured temperature of 0.98^0 C.

The high frequency short wavelength absorbed energy from the sun heats the earth, which in turn gives off Infrared energy that is longer wavelength and lower frequency heat energy in the Infrared Spectrum. As shown in Chapter 7 RF-4C IR Night Recon in Vietnam, a reconnaissance system sensitive to the IR spectrum can record high resolution television like pictures at night.

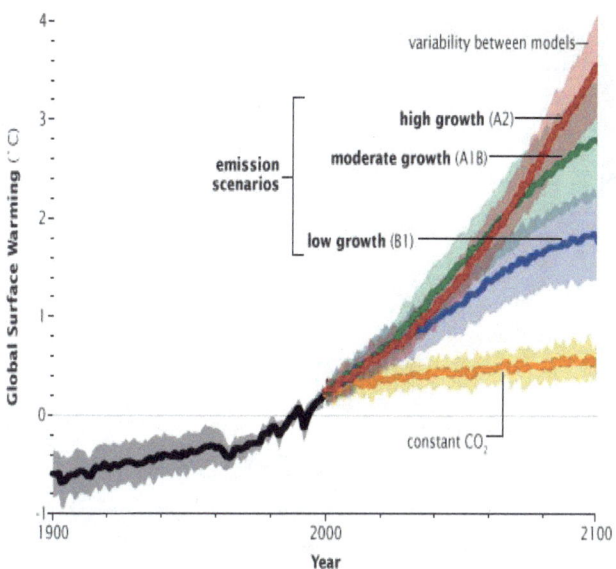

Figure 35.1 Temperature increase by Centuries (1900 – 2000 – 2100).

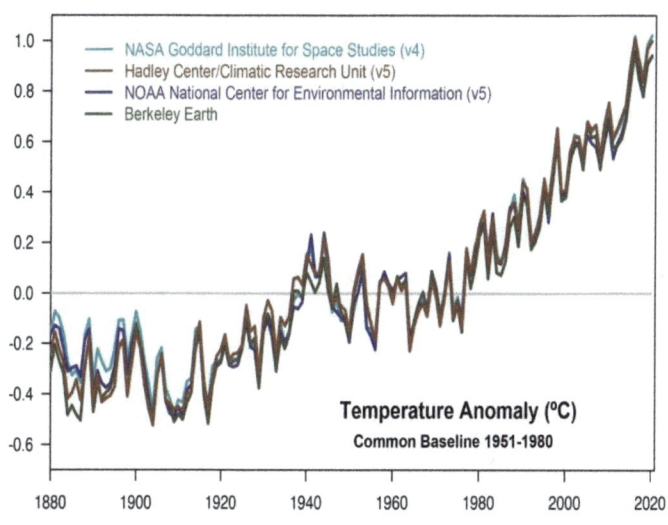

Figure 35.2 NASA Measured temperature increase comparisons.

Figure 35.3 Jacksonville, Florida Sunset in December 2020. Courtesy Norm & Denyse Sachman, our snowbird friends.

The winters in northern Florida are great with temperatures from 46°F to 65°F. The summers are hot with temperatures ranging from 75°F to 90+°F. With global warming the temperatures could increase in the future. Earth's global average surface temperature in 2020 tied with 2016 as the warmest year on record, according to an analysis by NASA. Continuing the planet's long-term warming trend, the year's globally averaged temperature was 1.84 degrees Fahrenheit (1.02 degrees Celsius) warmer than the baseline 1951-1980 mean, according to scientists at NASA's Goddard Institute for Space Studies (GISS) in New York.

According to an ongoing temperature analysis conducted by scientists at NASA's Goddard Institute for Space Studies (GISS), the average global temperature on Earth has increased by a little more than 1° Celsius (2° Fahrenheit) since 1880. Two-thirds of the warming has occurred since 1975, at a rate of roughly 0.15-0.20°C per decade.

Reconnaissance and ENERGY VISUALIZATION

by Stephen Weber

Part IV. Complex Systems and Problems.

Chapter 36. The weather.

How do we visualize energy? We visualize dynamic energy in our atmosphere via the weather. We see the effects of lightening at the speed of light lingering in the night sky. Many photographers have taken pictures of lightening. While the lightening discharge may appear to last seconds, there are many aspects of wind and water erosion that take hundreds or thousands of years to appreciate and understand such as the Grand Canyon. There are many aspects of the weather that we can watch in real time such as hurricanes and tornados which have tremendous amounts of energy and can be quite devastating. Pilots, mariners and people that work outside are concerned with the weather every day. That is why the weather forecast is always in the news. One way artists can depict energy is in the sky, whether it is sunny, cloudy or dramatic.

In 1989 I wrote a SBIR proposal to NASA for Satellite Weather Forecasting, Simulation and Control. It was under the distributed data base requirement. I proposed remote weather stations approximated to a grid and connected to a computer network. Each weather station would have the temperature, barometric pressure, the humidity, a rain gauge, the wind velocity and direction. The weather information from each weather station approximated to a grid, say every 100 miles, across our country roughly 3,000 miles across and 1200 miles down would imply 30 X 12 or 360 stations connected in a real time network. The data from each station could be overlaid on a map of our country and compared to a satellite weather map to provide detailed information of fronts for weather forecasting, prediction and modeling. The proposal had to be 25 pages long and 5 copies had to submitted to NASA.

Several months after submission I wrote a letter to Senator Paul Simon asking what happened to my proposal to NASA? Several weeks later I received his response that proposals from only 3 universities in the Big 10 were approved for a $50,000. SBIR grant. That was tough competition.

What causes the weather? The weather occurs in the earth's atmosphere. The atmosphere contains 6 layers and is approximately 60 miles high from the surface of the earth. The atmospheric air is more dense closer to the earth and less dense further away from the earth. The sun provides warmth to the earth, more in the summer and less in the winter because of the earth's tilt on its axis. At the same time the earth is rotating on its axis, once per day. There are 4 seasons: winter, spring, summer and fall. Two thirds of the earth's surface is water. There is ample opportunity for evaporation to take place and form clouds of water vapor. Rain occurs when the clouds are saturated with moisture. The solar radiation, the rotation and tilt of the earth on its axis and the ocean currents and wind all contribute to the weather. Earth's atmosphere has six layers: the troposphere, the stratosphere, the mesosphere, the thermosphere, the ionosphere, and the exosphere. The atmosphere of Earth protects life on Earth by creating pressure allowing for liquid water to exist on the Earth's surface, absorbing ultraviolet solar radiation, warming the surface through heat retention (greenhouse effect), and reducing temperature extremes between day and night. The greenhouse effect is caused by an increase in carbon dioxide from exhaust gases from automobiles, airplanes and coal power plants resulting in global warming.

The composition of the air in the atmosphere is 78% nitrogen and 21% oxygen according to NASA. There is only 0.93% argon and unbelievably only 0.04% carbon dioxide because the green plants absorb carbon dioxide and give off oxygen. There are trace amounts of helium, neon, methane, hydrogen and krypton. Most of the gases exist in the troposphere which is the closest layer to the earth's surface and extends 4 to 8 miles above the surface. Nearly all of the water vapor, dust and clouds are in this layer. This is where the dynamic weather occurs. The beautiful skies occur in the troposphere. The photographers and artists create pictures of skies. There is no distinct line between the earth's layers. The air gets colder as the altitude increases.

The next or second layer is the stratosphere which starts above the troposphere and ends about 31 miles or 50km high above ground. Ozone is plentiful in this layer and it heats up the atmosphere while absorbing harmful radiation from the sun. This air is very dry and thin and is where the jet aircraft fly. The next or third layer is the mesosphere which starts at 31 miles and extends to 53 miles above the ground and is the coldest layer with temperatures about 130 degrees below zero.

Meteors burn up in this layer. The thermosphere is the next layer and extends from the mesosphere at 53 miles to 62 miles or 100km above the earth. This is where the space shuttle and the space station orbit the earth. This is where charged particles from space collide with atoms and molecules and exhibit higher states of energy. This is where the auroras show their colors. The atoms with higher states of energy emit photons of light which we see as auroras. The next layer is the exosphere, which is the highest and thinnest layer and is composed of particles of hydrogen and helium and is where the atmosphere merges with space.

Meteorologists can measure the basic elements of the weather which are: temperature, wind direction and velocity, atmospheric or barometric pressure, humidity, precipitation-rain or snow, and cloudiness, partly cloudy or sunny. What causes the weather? The weather occurs in our atmosphere.

The earth orbits around the sun once a year, while it rotates on its axis which is tilted once per day. There are four seasons: spring, summer, fall and winter which influence the weather. Some say that the weather occurs over the oceans since the earth rotates on its axis. What about La Nina and El Niño? What about the hurricanes from Africa that affect Florida? What affect do greenhouse gasses and global warming have on the weather? Chicago is called the Windy City. They say, "If you don't like the weather, wait a day and it will change."

Reconnaissance and ENERGY VISUALIZATION

By Stephen Weber

Part IV. Complex Systems and Problems.

Chapter 37. Energy in Transportation.

For several thousand years wind energy has powered sailing ships on the lakes and oceans around the world. James Watt worked on developing an improved steam engine from 1763-1775 to provide rotary motion and consequently accelerated the Industrial Revolution that led to steam boats and steam locomotives. There were many developments to improve the efficiency of steam engines in the 18th and 19th century. Steam locomotives led to the expansion of the railroad across the United States in the 19th century. The steam locomotives of the 19th and early 20th century have been replaced by the diesel locomotives. The diesel engine charges the battery that powers the DC motor to drive the wheels of the trains. This is an efficient hybrid technology.

With the development of the gasoline powered internal combustion engine in the 20th Century, there has been a tremendous increase in the number, types and design of automobiles in the past 120 years. There are luxury sedans, trucks, sports cars, and SUVs with V8s, V6s and 4 cylinder economy engines. With the threat of global warming and the fluctuating price of gasoline, hybrid vehicles such as the Prius, which gets over 50 mpg, and lithium battery powered vehicles such as the Tesla, which gets over 300 miles per charge, have gained in popularity.

In the aircraft industry, the internal combustion engine reached its peak in WWII where over 20,000 B-24 propeller aircraft were manufactured and approximately over 15,000 B-17 aircraft were manufactured in the United States. After WWII in the Korean War jet engines became more popular because of their speed. Today almost all commercial airlines use jet powered aircraft. With the advent of space technology, rocket powered engines which use hydrogen are used with the space shuttle and to go to the moon and back with the Apollo series in the 1960's.

The first generation of Aircraft used gas engines to power propellers. There were propeller driven single engine private aircraft and propeller driven multi engine military aircraft such as the B-17 and B-24 used in WWII. There were commercial propeller aircraft. These were replaced with second generation jet engine commercial aircraft which were typically multi engine design that influenced the entire transportation industry as well as jet engines for military superiority. The third generation aircraft will be battery powered electric aircraft. NOVA had several one hour shows on electrical aircraft. The first was a solar powered French aircraft that flew around the world. The second show was on battery powered propeller aircraft with electrical motors to propel the aircraft. They had both single engine and push pull multi engine electric aircraft.

With Navy aircraft carriers and submarines, nuclear technology is used for propulsion systems as well as Edison nuclear power plants. With the advent of battery powered vehicles, there is the necessity of recharging these batteries daily from our power plants. Through the 1950's coal powered electric power plants were very common and the coal dust covered the cityscapes in the winter. Hydroelectric, natural gas, wind, and solar are the more common electric power plants.

In 2020 there were 282 million cars in the United States. There were 1.4 billion cars in the world. Each year 100 million new cars are manufactured. President Biden wants to build 50,000 new charging stations in the United States to recharge all the new battery powered vehicles that are planned. The problem is that this will put a burden on the existing fossil fuel (coal, gas and oil) electric power plants in our country.

In order to reduce global warming, we need to reduce the amount of carbon dioxide in the atmosphere. A conventional internal combustion engine inputs oxygen and gasoline and exhausts carbon dioxide. We need an engine that operates like a tree and takes in carbon dioxide and exhausts oxygen. The best way to remove carbon dioxide from the atmosphere is to plant trees and stop deforestation and stop forest fires. We need engines that function like trees to remove the carbon dioxide and give off oxygen. Perhaps we need a design competition to create a photosynthesis clean air engine that provides power by running on carbon dioxide in the air and exhausting oxygen as shown in figure 37.1.

Figure 37.1 The Green Clean Air Engine.

The green clean air engine functions like a tree and absorbs Carbon dioxide and exhausts oxygen, thus cleaning our atmosphere. This is a necessary reverse hypothetical concept.

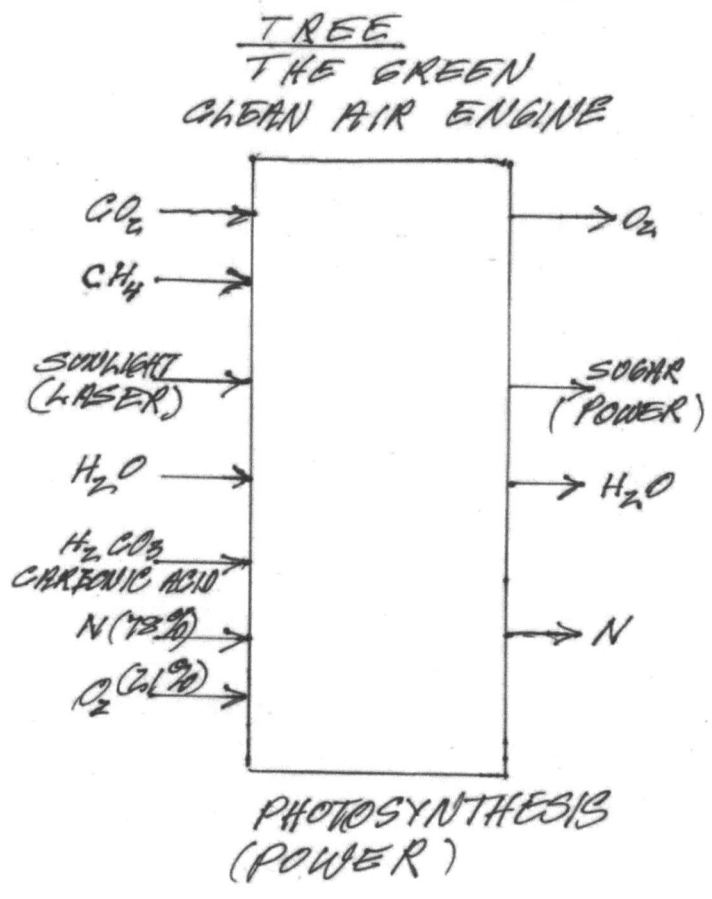

Figure 37.2 The Green Clean Air Machine Block Diagram.

I was asked why not have a battery powered electric vehicle that can recharge itself from braking and photo cells? Universities have solar powered electric cars that comepete in trips across America. These are one man lightweight aero dynamic experimental sports cars. There was a French solar powered plane with electric engines that flew around the world. A battery powered vehicle could have a small gas engine to charge the battery for continuous operation.

Reconnaissance and ENERGY VISUALIZATION
By Stephen Weber

Part IV. Complex Systems and Problems.

Chapter 38. Global Warming.

Ever since the movie: *"An Inconvenient Truth."* 2006, presented in film an illustrated talk on climate by former Vice President Al Gore, the world has been concerned about global warming and climate change. Al Gore shows in the movie how the glaciers and polar ice caps have been melting due to the emission of green house gasses. This could cause a rise in ocean levels and affect all the cities and states bordering the oceans such as New York City and Florida. The increase in temperature is expected to be between 2^0C and 6^0C by the end of the 21st century.

There are many questions related to global warming that should be answered for understanding:
1. How long has global warming been going on? Since the last ice age or more realistically since WWII and atomic and hydrogen bomb testing have increased thermodynamic entropy.
2. Does global warming lead to climate change? Yes. Climate change is the physical evidence that global warming is slowly melting glaciers and polar ice caps.
3. Is global warming reversible? Yes. It depends on time and at what point on the exponential temperature curve the earth is at when serious reversal efforts begin. The goal of the Paris Climate Accord is zero carbon emissions by the year 2050.
4. What causes global warming? Short wave radiation from the sun 93 million miles away warms the earth which gives off long wave infrared IR radiation. The greenhouse gasses in the lower atmosphere reflect the infrared energy back down toward earth.

5. What are the greenhouse gasses in their order of importance? The worst of the greenhouse gasses is water vapor H_2O, because as the atmosphere is heated it can hold more moisture. The next worse is methane CH_4, because as the earth's permafrost heats up more methane is released which is 10 times more dangerous than carbon dioxide. The next worse is carbon dioxide CO_2, because cars and animals give off carbon dioxide, whereas trees take in carbon dioxide and give off oxygen cleaning our air. Nitrous oxide N_2O and ozone O_3 are additional trace amount greenhouse gasses.
6. Why is water vapor considered a greenhouse gas? Infrared IR radiation is hygroscopic; it is absorbed by the clouds. When our RF-4C Air Force recon pilots flew over clouds with the IR system on, you could see the clouds distinctly just like optical photography. Whereas, radar could go through the clouds. The IR radiation is absorbed by water vapor.
7. Why is the infrared radiation absorbed by the greenhouse gases and the water vapor? When the IR radiation hits a CO_2 molecule, it absorbs the energy and starts vibrating raising the temperature of the gas. Then the molecule of CO_2 releases an IR photon towards earth and stops vibrating. The gas is warmed and the earth is warmed. As the air temperature is increased the humidity increases, because warm air can hold more moisture or water vapor than cold air.
8. What phenomena are associated with climate change? We recognize the seriousness of global warming because of the visual effects of climate change. These are: average annual temperature increase of several degrees resulting in polar ice caps starting to melt, glaciers melting, ocean temperature increase and sea level increase. Warm air can hold more moisture so the severity and frequency of the storms can also increase.

The amount of radiation emitted by an object depends on the temperature of the object. The hotter, the greater the IR radiation! This relationship is the Stefan-Boltzmann law. $J = \alpha T^4$, where J= joules /sec /square meter or energy per unit time per unit area. α = Stefan Boltzmann constant = 5.670373×110^{-8} watts / m^2 K^4, where K is the temperature in Kelvin. This Stefan Boltzmann law explains how much infrared energy the earth will emit per unit area.

As sunlight warms the earth there is an energy transformation. High frequency light energy is changed to thermal heat energy and low frequency infrared radiation. The greenhouse gasses, which comprise less than 1.0% of the earth's atmosphere, absorb some of the infrared radiation produced by Earth's surface. 78% of our atmosphere is nitrogen and 21% oxygen. Because greenhouse gases emit the same amount of radiation they absorb and because this radiation is emitted equally in all directions, the net effect of absorption by greenhouse gases is to increase the total amount of radiation emitted downward toward Earth's surface.

The earth's surface is warmed by the Greenhouse gasses by increasing the net downward infrared radiation reaching the earth's surface. The relationship between atmospheric concentration of greenhouse gases and the associated downward radiation force toward the surface is different for each gas. The main greenhouse gasses are: CO_2, CH_4, and $H_2 0$. A complicated relationship exists between the chemical properties of each greenhouse gas and the relative amount of long wave infrared radiation that each can absorb.

Water vapor $H_2 0$ is the most potent of the greenhouse gases in Earth's atmosphere, because infrared energy is hygroscopic and absorbed by clouds, therefore its behavior is fundamentally different from that of the other greenhouse gases. The primary role of water vapor is a climate feedback as a response within the climate system that influences the system's continued activity. This distinction arises from the fact that the amount of water vapor in the atmosphere is a function of the warmer the surface, the greater the evaporation rate of water from the surface. As a result, increased evaporation leads to a greater concentration of water vapor in the lower atmosphere capable of absorbing infrared radiation and emitting it downward.

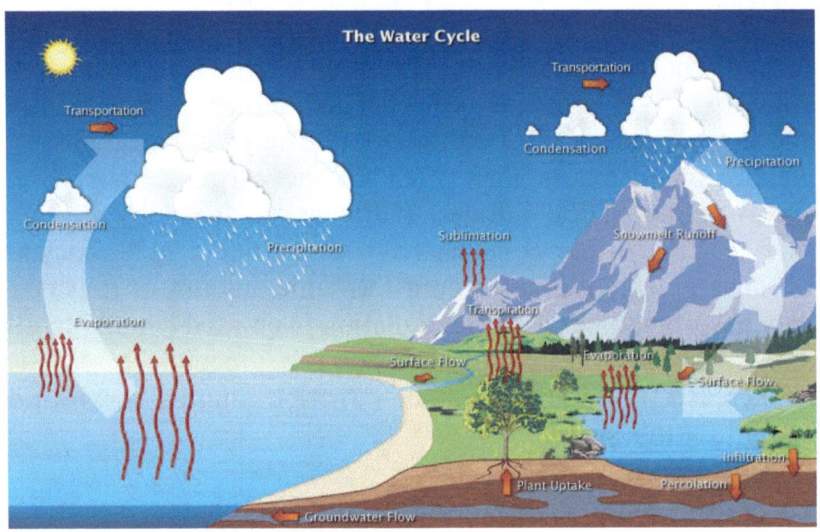

Figure 38.1 The Water Cycle. Courtesy NASA.gov

The contemporary water cycle shows water evaporation from the oceans into clouds in the atmosphere and through transportation, condensation, sublimation and precipitation to the continents and rivers back to the oceans again. As the earth warms up, the atmosphere can hold more moisture and the severity of the storms, floods and hurricanes may increase. An increase of water vapor in the atmosphere makes it more difficult for the IR radiation to escape because it is hygroscopic and absorbed by the water vapor in the clouds.

Carbon dioxide (CO_2) comprises only 0.04% of the atmosphere. Natural sources of atmospheric CO_2 include active volcanoes the combustion and natural decay of organic matter, automobiles, and respiration by oxygen using organisms. These sources are balanced by processes, called "sinks," that tend to remove CO_2 from the atmosphere. These include green vegetation, trees and grass which absorb CO_2 during the process of photosynthesis and give off oxygen.

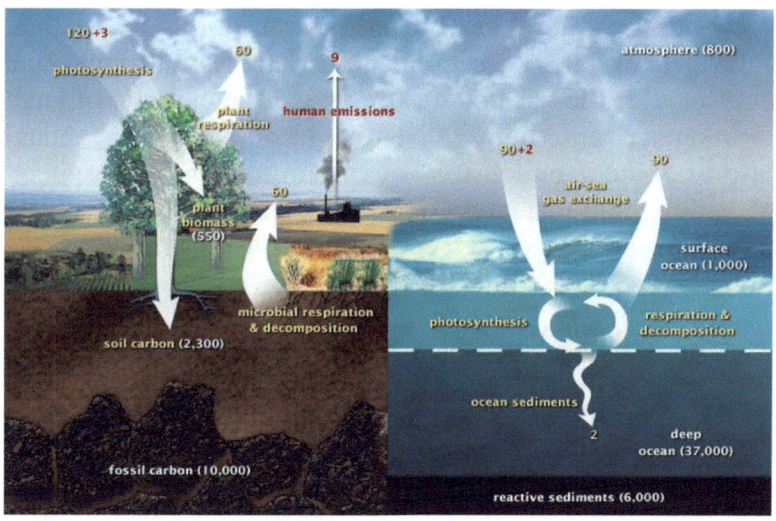

Figure 38.2 Fast Carbon Cycle. Courtesy NASA.gov

Carbon is transported in various forms through the atmosphere, the hydrosphere, and geologic formations. Acid Rain, the primary pathway for the exchange of carbon dioxide (CO_2) takes place between the atmosphere and the oceans; there a fraction of the CO_2 combines with water, forming carbonic acid (H_2CO_3) that subsequently loses hydrogen ions (H^+) to form bicarbonate (HCO_3^-) and carbonate (CO_3^{2-}) ions.

Carbon dioxide also exchanges through photosynthesis in plants and through respiration in animals. Dead and decaying organic matter may ferment and release CO_2 or methane (CH_4) or may be incorporated into sedimentary rock, where it is converted to fossil fuels. Burning of hydrocarbon fuels returns

CO_2 and water (H_2O) to the atmosphere. The biological and anthropogenic pathways are much faster than the geochemical pathways and, consequently, have a greater impact on the composition and temperature of the atmosphere.

Human activities increase atmospheric CO_2 levels primarily through the burning of fossil fuels—principally oil and coal and secondarily natural gas, for use in transportation, heating, and the generation of electrical power. Other anthropogenic sources include the burning of forests and the clearing of land. Anthropogenic emissions currently account for the annual release of about 7 billion tons of carbon into the atmosphere. As the population increases, CO_2 emissions increase. As trees are cut down or burned, CO_2 emissions are increased.

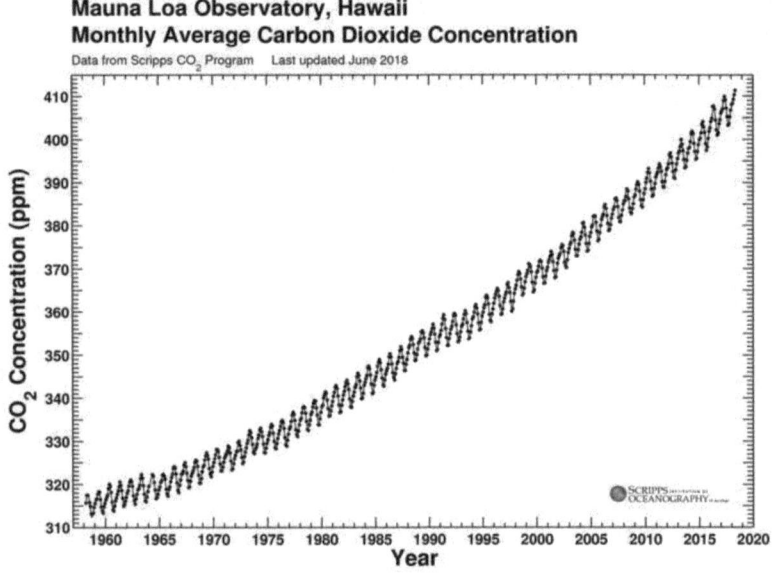

Figure 38.3 Scripps Keeling Curve. National Geographic.

The Keeling Curve tracks changes in the concentration of carbon dioxide (CO_2) in Earth's atmosphere at a research station on Mauna Loa in Hawaii. The overall trend shows that CO_2 is increasing in the atmosphere.

According to Encyclopedia Britannica and Britannica Online Encyclopedia the surface warming potential will rise by roughly the same amount for each doubling of CO_2 concentration. At current rates of fossil fuel use, a doubling of CO_2 concentrations over preindustrial levels is expected to take place by the middle of the 21st century A doubling of CO_2 concentrations would represent an increase of roughly 4 watts per square meter of radiation forcing. Given typical estimates of "climate sensitivity" in the absence of any offsetting factors, this energy increase would lead to a warming of 2 to 5 °C by the end of the 21st century.

The second most important greenhouse gas is Methane. CH_4 is more potent than CO_2 because the radiation forcing produced per molecule is greater. In addition, the infrared window is less saturated in the range of wavelengths of radiation absorbed by CH_4, so more molecules may fill in the region. However, CH_4 exists in far lower concentrations than CO_2 in the atmosphere, and its concentrations by volume in the atmosphere are generally measured in parts per billion rather than parts per million. CH_4 also has a considerably shorter residence time in the atmosphere than CO_2.

Sources of methane include tropical and northern wetlands, methane-oxidizing bacteria that feed on organic material consumed by termites, volcanoes, seepage vents of the seafloor in regions rich with organic sediment, and methane hydrates trapped along the continental shelves of the oceans and in polar permafrost. The primary natural sink for methane is the atmosphere and the tundra permafrost.

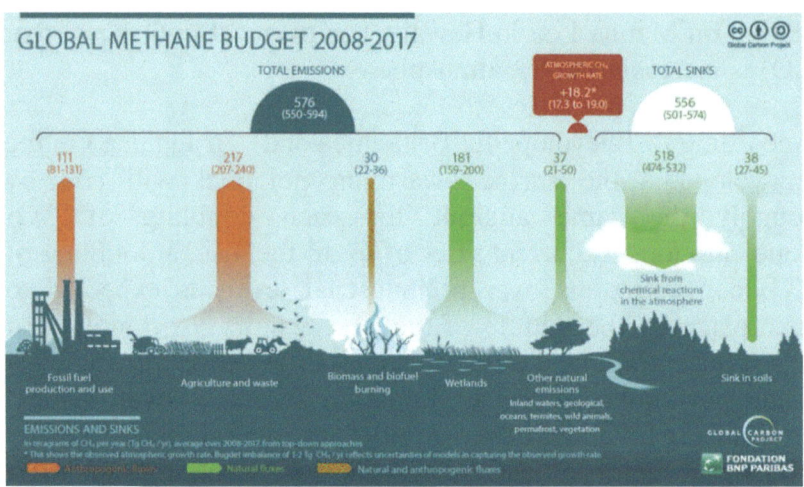

Figure 38.4 Main Sources of Methane. Courtesy Wikipedia.

Human activity is increasing the methane CH_4 concentration faster than it can be offset by natural sinks. The major anthropogenic sources of atmospheric CH_4 are rice cultivation, livestock farming, the burning of coal and natural gas, the combustion of biomass, the decomposition of organic matter in landfills and the permafrost.

The global carbon cycle involves the two main reservoirs of carbon in the climate system, the oceans and the terrestrial biosphere. These reservoirs have historically taken up large amounts of anthropogenic CO_2 emissions. Global warming, however, could reduce the rate of carbon uptake by these reservoirs and would increase the pace of CO_2 buildup in the atmosphere and represent increased greenhouse gas.

As surface waters warm, the oceans would hold less dissolved CO_2. If more CO_2 were added to the atmosphere and taken up by the oceans, bicarbonate ions (HCO_3^-) would multiply and ocean acidity would increase (Acid Rain). Since calcium carbonate ($CaCO_3$) is broken down by acidic solutions, rising acidity may affect fauna that use $CaCO_3$ into their skeletons.

According to Encyclopedia Britannica and Britannica Online Encyclopedia: rising surface temperatures might lead to a slowdown in thermo circulation, a global pattern of oceanic flow that partly drives the sinking of surface waters near the poles and is responsible for much of the burial of carbon in the deep ocean. A slowdown in this flow due to an influx of melting fresh water into what are normally saltwater conditions might also cause the solubility pump, which transfers CO_2 from shallow to deeper waters, to become less efficient. It is predicted that if global warming continued to a certain point, the oceans would cease to be a net sink of CO_2 and would become a net source.

If the Amazon rain forest is lost because of the global warming, the overall capacity of plants to sequester atmospheric CO_2 would be reduced. As a result, the terrestrial biosphere, though currently a carbon sink, would become a carbon source. Ambient temperature is a significant factor affecting the pace of photosynthesis in plants, and many plant species that are well adapted to their local climatic conditions have maximized their photosynthetic rates. As temperatures increase and conditions begin to exceed the optimal temperature range for both photosynthesis and soil respiration, the rate of photosynthesis would decline.

There is also the potential for increased methane release as a result of the warming of Arctic permafrost (on land) and further methane release at the continental margins of the oceans. There are at least 400 billion tons of carbon equivalent stored in Arctic permafrost and as much as 10 trillion tons of carbon equivalent trapped on the continental margins. Methane is 21 times more potent than carbon dioxide.

Computer models are use for weather forecasting and climate modeling. They simulate the responses and interactions of the oceans and atmosphere, and to account for changes to the land surface, both natural and human-induced. They comply with the laws of physics, conservation of energy, mass, momentum and account for many factors that influence earth's climate.

The models are complicated and rigorous tests with real-world data hone them into powerful tools that allow scientists to explore our understanding of climate in ways not otherwise possible. By experimenting with the models—removing greenhouse gases emitted by the burning of fossil fuels or changing the intensity of the Sun to see how each influences the climate—scientists use the models to better understand Earth's current climate and to predict future climate changes.

According to the climate model simulations, as the world consumes ever more fossil fuel, greenhouse gas concentrations will continue to increase, and Earth's average surface temperature will rise with them. Based on a range of plausible emission scenarios, average surface temperatures could rise between 2°C and 6°C by the end of the 21st century.

In 1986 I did computer simulations with IBM's GPSS (General Purpose Simulation Software) at UIC's graduate college of Engineering on an IBM 3081 running MVS for UIC's LCS (Library Computer System). Much of the accuracy of the simulation depends on design of the computer model.

In the past several years the forest fires in the western United States have been disastrous. One whole town in the path of the forest fire, Paradise, California burnt to the ground. Then there was a shortage of water in California. In the southern hemisphere, the Australian continent had forest fires and the Amazon rain forest is being deforested.

They are cutting down the trees that absorb carbon dioxide and give off oxygen. These trees give off pollen dust and the water vapor molecules congeal on the pollen dust and this forms a moisture river in the atmosphere above the trees in the Amazon rain forest. One tree in this rain forest consumes 250 gallons of water per day. Moisture droplets attach to the pollen dust particles creating a cloud or moisture rivers in the atmosphere above the trees. If you multiply this by millions of trees, you have a rain forest.

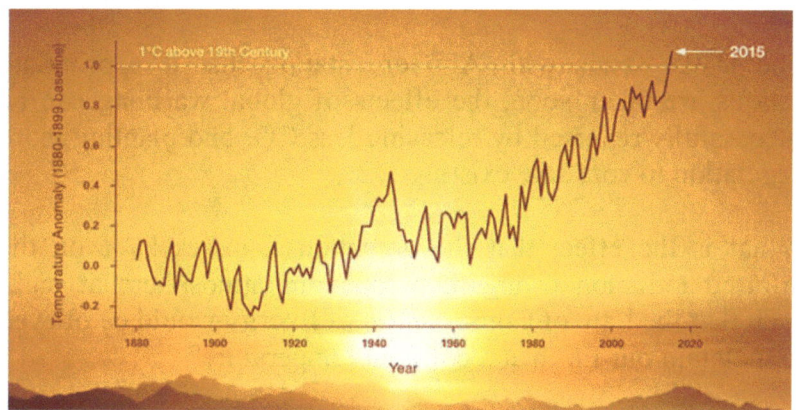

Figure 38.5 Global Temperature. Courtesy NASA Cal Tech.

According to Al Gore, "An Inconvenient Truth" presents in film form an illustrated talk on climate by Al Gore, aimed at alerting the public to an increasing planetary emergency due to global warming. I bought Al Gore's DVD An Inconvenient Truth to show to my students to help make science interesting

Al Gore begins his slide show on Global Warming; a comprehensive presentation with detailed graphs, flow charts and stark visuals. Al Gore shows off several photographs of the Earth taken from multiple space missions. Al Gore notes that these photos dramatically transformed the way we see the Earth, helping spark modern environmentalism.

Al Gore discusses the scientific opinion on global warming, as well as the present and future effects that global warming will produce if the amount of human generated greenhouse gases are not significantly reduced in the very near future. Gore also presents Antarctic ice coring data showing CO_2 levels higher now than in the past.

The film includes segments intended to refute critics who say that global warming is unproven or that warming will be insignificant. For example, Al Gore discusses the possibility of the collapse of a major ice sheet in Greenland or in West Antarctica, either of which could raise global sea levels.

The DVD ends with Al Gore stating that if appropriate actions are taken soon, the effects of global warming can be successfully reversed by releasing less CO_2 and planting more vegetation to consume existing CO_2.

What is the effect that the greenhouse gasses have on the infrared radiation trying to escape our atmosphere at night such that the Law of Conservation of Energy would be obeyed to maintain our Earth at a constant temperature?

Figure 38.6 "An Inconvenient Truth." DVD by Al Gore.

The main greenhouse gasses are carbon dioxide (CO_2) and methane (CH_4). The political goal of the climate scientists has been to reduce the amount of carbon emissions into our atmosphere to reduce global warming by reducing coal and fossil fuel power plants and require much higher miles per gallon standards for automobiles or even battery powered cars.

Since radiation from the sun is reflected off the clouds in our atmosphere, another approach is to create more clouds by atomizing water vapor with reflective properties into our atmosphere. This is what happened millions of years ago when a large asteroid hit the earth and created such a huge dust cloud and the earth was plunged into an ice age.

The world has just 12 years in which to cut emissions by half, and 32 years (by 2050) to get net emissions down to zero. However, far from the required dramatic decline, the latest figures show global carbon dioxide emissions actually going up. And some think that artificially engineered reflective clouds may be one important solution.

Currently around 30% of the sun's rays that reach Earth are reflected back to space by white surfaces, largely our polar ice. Sea ice reflects sunlight better than any known natural surface, bouncing around 90% back up; at the other end of the scale, the dark open ocean reflects just 6% of sunlight and absorbs 94%. With Arctic sea ice rapidly declining perhaps we can expect to see global warming accelerate.

Another approach mentioned was to plant hundreds of millions more trees, because the leaves on trees in the photosynthesis process absorb carbon dioxide and give off oxygen. The IR radiation continually bumps into carbon atoms in the process of trying to leave our atmosphere and ends up warming our atmosphere.

In summary, everything possible must be used together to reduce global warming. All of these approaches must be used in combination to reduce global warming and climate change. The Global Temperature Anomaly Figure 38.5 Shows an increase in temperature during WWII with aircraft bombing and especially the atomic bombs: Little Boy and Fat Man. In order to reduce global warming and climate change, it is imperative that we end war and work for peace in the world.

Figure 38.7 David Attenborough, Climate Change- The Facts. PBS Television Special 2020.

According to David Attenborough there is still time. Urgent action is needed. What can be done? We must take dramatic action in the next decade. Twenty of the last 22 years had record high temperatures. Coal, gas, and oil, all fossil fuels, when burnt give off carbon dioxide, which was at 220 ppm now is at 440 ppm, In 100 years CO_2 emissions have doubled. The global temperature has gone up 1°C. The frequency of extreme temperatures is increasing. 8% of species are at the threat of extinction. What does 1°C mean? Last year we saw record wild fires in Paradise, California and in the western United States. Global Warming is changing the weather.

More heat means more moisture in the atmosphere, more rains and more flooding. The Greenland ice sheet is melting. Antarctica is losing 3 times as much ice. The sea level is going up. It has risen 20 cm in the last 20 years. In the US, Louisiana is losing land. There are climate refugees. 90% of the increased heat is stored in the oceans. Adaptation is not a solution. The Land sat satellite shows that the world's great forests like the Amazon rain forest are being deforested for palm oil. One third of CO_2 emissions are caused by deforestation. Looking ahead, increased CO_2 emissions are much worse. Numerical model simulations predict 1 ½ °C increase in temperature by 2040 to 2050. Climate projections predict a 3 °C to 6 °C increase by this century's end in 2200 and an 80cm increase in the sea level.

Major ECO systems could turn against us. Permafrost with methane is 21 times worse than CO_2. There is still time if we act now. The 2015 climate summit in Paris proposed to limit the increase to 1 ½ °C and shift from fossil fuels to renewable energy sources. Everything has a carbon footprint. We can be less wasteful. In Australia the school kids had a school strike for climate change. The kids will inherit the earth.

David Attenborough was preparing for a Glasgow, Scotland Climate Change Summit and said the important thing was to get China on board and involved with climate change. David Attenborough is a world famous naturalist and his opinions are highly valued. According to David Attenborough there is still time to solve the climate crisis but we must start now. David Attenborough is an optimist. The science says that global warming is a real process and according to the second law of thermodynamics a real process is irreversible.

Reconnaissance and ENERGY VISUALIZATION

By Stephen Weber

Part IV. Complex Systems and Problems.

Chapter 39. Entropy and Thermodynamics.

Entropy is the degree of disarray in a system. As the heat in a system increases, the entropy increases. Entropy is an increasing function. Entropy is a thermodynamic quantity representing the unavailability of a systems thermal energy for conversion into mechanical work often interpreted as the degree of disorder or randomness of the system. Entropy is a measure of a systems thermal energy per unit temperature that is unavailable for doing useful work. S = Entropy of a system.

$dS = \dfrac{\delta Q}{T}$, where dS= infinitesimal increment of Entropy, δQ = infintesimal transfer of heat, T = common temperature.

The first law of thermodynamics expresses the law of conservation of energy and provides the definition of the internal energy of a system.

The second law of thermodynamics is concerned with the direction of natural processes. The second law asserts that a real or natural process runs only in one sense or direction and is not reversible.

By definition entropy S is a positive increasing function therefore:

$$\oint \dfrac{\delta Q}{T} \geq 0, \quad \oint \dfrac{\delta Q}{T} = S, \quad \dfrac{dS}{dt} = S_i \geq 0.$$

For example heat always flows from a hotter to a colder body. Another global warming example: Consider a glacier slowly moving down a mountain to the sea. As the temperature increase over time, the glacier moves faster. The process is not reversible. Even if there is a cold winter, the glacier cannot flow backwards up the hill. The process is not reversible. Consider large chunks of ice falling off Antarctica or Greenland into the ocean. These processes are not reversible. The chunks of fallen ice cannot rise up and attach themselves to Antartica or Greenland. The physical effects of global warming are not reversible but can be slowed down.

President Joe Biden appointed John Kerry US Presidential Envoy for Climate Change. President Joe Biden rejoined the Paris Climate Change Agreement. Former Secretary of State John Kerry said the United States will work on 3 fronts to promote ambition, resilience and adaptation at the inaugural Climate Adaptation Summit. According to John Kerry the US will leverage innovation and climate data and information to promote a better understanding of climate risk.

Secondly, the US will significantly increase the flow of finance through adaptation and resilience initiatives. Third the US will partner with the private sector domestically and in developing countries to promote a greater collaboration between business and the communities in which they serve.

We have learned that green trees and plants absorb carbon dioxide and give off oxygen in the process of Photosynthesis. We have also learned that AGENT ORANGE defoliant kills green plants, trees, grass and people and subsequently causes climate change and global warming.

We have also learned that war causes an increase in global warming and climate change. According to Figure 38.5 the Global Temperature Anomaly from NASA at Cal Tech JPL there is a big spike in temperature during and after WWII with the use of atomic bombs and the subsequent testing of large scale hydrogen bombs after WWII and the Korean War. We must work for peace in the world to mitigate the climate crisis.

In summary, the hypothesis of this book is that we can see energy. In the visible light spectrum we see light. Light is energy as shown in Chapter 12. The French Impressionists painted light as shown by Claude Monet's beautiful paintings in Chapter 22. What about the other significant areas of the electromagnetic spectrum?

The X-Ray spectrum is much higher in frequency than light and shorter in wavelength, but in Chapter 33 Medical X-Ray we can use X-Ray to view inside the human body. We can see the shadow of bones that absorb the energy but the X-Rays pass through the soft tissue and expose the film. We can see the results of the X-Ray energy.

The infrared spectrum is adjacent to and larger than the light spectrum. How can we see infrared energy? In Part I, Chapter 4, The RF-4C Infrared Reconnaissance System is used by the Air Force to view everything at night without light. We are viewing thermal energy. The hotter an object the more IR infrared energy is given off. This is what is peculiar about global warming. According to the balance of energy, this IR radiation would normally escape from our atmosphere. But because there are carbon dioxide emissions from cars and methane emissions from permafrost melting, the IR energy is reflected back down to earth and increases the temperature resulting in global warming and climate change.

Satellite reconnaissance was used by Al Gore in his DVD to show that the polar ice caps are melting as proof of climate change. When high frequency sunlight warms the earth, there is an energy transformation and heat energy and lower frequency infrared radiation is given off according to the Stefan-Boltzmann law. Our President Joe Biden and Secretary

John Kerry rejoined the Paris Climate Change Agreement because it will take a global effort to combat global warming and climate change.

In summary, Entropy is an increasing function. The 2nd Law of Thermodynamics states that a real process flows in only one direction and cannot be reversed. Global warming is a real process and is irreversible. But, how much can global warming be slowed down? Each year the average temperature is higher than the previous year. The average temperature for the last ten years is usually higher than the average temperature for the previous ten years. The average temperature of the ocean is increasing. The world population is increasing. There are 7.9 billion people on earth. This alone increases the entropy and the average temperature because humans inhale oxygen and exhale carbon dioxide.

Aerial and space reconnaissance has captured the visual effects of climate change. The Greenland and Antarctic ice has decreased in mass. Greenland lost an average of 279 billion tons of ice per year between 1993 and 2019 according to NASA's Gravity Recovery and Climate Experiment, while Antarctica lost about 148 billion tons of ice per year. Glaciers are retreating in the Alps, Andes, Rockies and Himalayas. According to the 2nd Law of Thermodynamics, a real process moves in one direction and is not reversible. Glaciers are not going back uphill and 279 billion tons of ice will not be added to Greenland. In fact, satellite recon has shown that the spring snow cover in the northern hemisphere has decreased in the past 50 years and the global sea level has risen about 8 inches in the past century and the rate in the past 20 years is nearly double the rate of the last century. The extent and thickness of the arctic sea ice has declined rapidly in the past 2 decades. In the United States the number of record high temperature events has been increasing resulting in record forest fires. The acidity of the ocean surface waters has increased by 30%, because of the increase in carbon dioxide emissions.

Reconnaissance and ENERGY VISUALIZATION

By Stephen Weber

Part IV. Complex Systems and Problems.

Chapter 40. James Webb Space Telescope and Dark Energy.

NASA's JWST, the successor to the Hubble telescope, was launched in December 2021. After a complex unfolding in space 1 million miles from earth, the first images were released by NASA in July 2022. The goal of JWST is to view back 13.8 billion years ago to the beginning of our universe just after the Big Bang to see how the stars were born and the galaxies were formed. JWST uses the infrared spectrum and can see the red shift of the stars in the expanding universe. The large sunshield and unfolding antenna had to fit into a nose cone 5 meters in diameter and then unfold in space I million miles from earth and operate in cryogenic temperatures 55 degrees above absolute zero. JWST was so large that it could not be tested on earth before launch. Only small portions could be tested and then mathematically simulated before launch. The images generated are high resolution beyond expectation.

The black and white infrared video signal is converted to color by assigning red to the longest wavelengths and then green to the medium wavelengths and blue to the shortest wavelengths. This is the standard RGB red, green, blue of color television. The antenna consists of 18 hexagonal mirrors or elements made of beryllium and plated with gold which reflects the IR energy. JWST has 4 infrared cameras and 2 IR spectrometers. JWST has a sunshield composed of 5 thin layers and is as large as a tennis court to protect the James Webb Telescope whose mission was to look far back to the beginning of the universe to see the first stars ever born just after the Big Bang explosion to capture the accelerating and expanding universe.

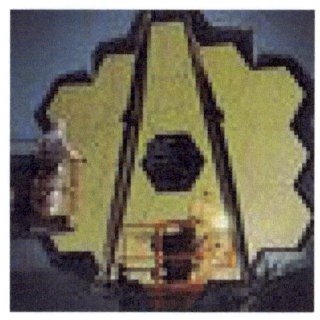 

Fig 40.1

The expanding universe implies Dark Energy which is by far the largest component of the space in our universe. There is 5 times more Dark Matter in our universe as there are stars and planets. Black holes are considered to be Dark Matter and are thought to be the center of every galaxy. Our Milky Way galaxy has a black hole as the center with everything revolving around the super massive black hole, 27,000 light years from earth, called Sagittarius A star.

The JWST is expected to play a significant role is researching Dark Energy, however, the IR cameras aboard JWST visualize infrared wavelengths and not electromagnetic or gravity waves. Perhaps the forces holding our expanding universe together would have to be calculated theoretically and proven mathematically. Perhaps black holes at the center of the galaxies are the dark matter which creates the dark energy in our universe. The James Webb Space Telescope is the greatest research vehicle to visualize both dark matter and dark energy.

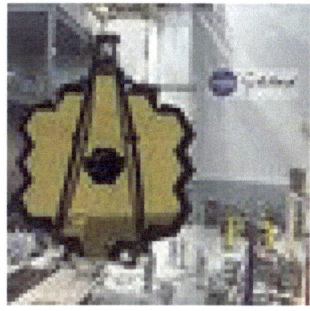 

Fig 40.2

Reconnaissance and ENERGY VISUALIZATION

By Stephen Weber

Part IV. Complex Systems and Problems.

Acknowledgements

Figure 35.1 Temperature Increase by Centuries.

Figure 35.2 NASA Measured temperature increase.

Figure 35.3 Jacksonville, Florida Sunset in December 2020.

Figure 37.1 The Green Clean Air Engine.

Figure 37.2 The Green Clean Air Machine Block Diagram.

Figure 38.1 The Water Cycle. Courtesy NASA.gov

Figure 38.2 Fast Carbon Cycle. Courtesy NASA.gov

Figure 38.3 Scripps Keeling Curve. National Geographic.

Figure 38.4 Main Sources of Methane. Courtesy Wikipedia.

Figure 38.5 Global Temperature. Courtesy. Cal Tech JPL.

Figure 38.6 "An Inconvenient Truth." DVD by Al Gore.

Figure 38.7 David Attenborough, Climate Change- The Facts, PBS Television Special. 2020.

Figure 40.1, 40.2 James Webb Space Telescope pictures. Courtesy NASA.gov.

Reconnaissance and ENERGY VISUALIZATION.

by Stephen E. Weber

APPENDIXES.

Appendix A. You Are Not Alone.

Edward Hines Jr. VA Medical Center Cancer Survivors' Celebration

Friday June 1, 2001 A Celebration of Life!

Speech: "**You Are Not Alone**." by Stephen Weber

Opening Prayer:

Dear Lord thank you for your mercy and allowing us to be here today. Thank you for our great doctors, nurses and hospital personnel. Dear heavenly father thank you for showing us that we are not alone. Thank you for the dedicated team of health professionals, our family and friends in helping us in our fight against cancer. Thank you for this opportunity to celebrate life together as survivors. Amen

Diagnosis:

A little over a year ago they said I had a gastric problem so they ordered an endoscopic test. That's when the price of unleaded gas was pushing $2.00 a gallon and the oil companies blamed OPEC. This summer the oil companies said that the old refineries can't keep up with the EPA's new requirements for 15 different grades of summer reformulated gasoline and that's just in the Midwest. Isn't that the old supply and demand price gouging theory?

Exactly one year ago the endoscopic biopsy and pathology report indicated gastric adeno carcinoma and I was told that I would have to have my stomach removed. My first thought was that I was alone and needed a second opinion because I thought I would starve to death without a stomach. My second thought was that the stock market was going downhill with me. The market started getting anemic in March of 2000 when my blood tests revealed that I was anemic. The stock values of the dot com companies that were not showing a profit really went down. Instead of taking my money out of the market, I made out my will and watched the stock prices.

General surgery set up an appointment with the chief of oncology, Dr. Bhoopalam, who ordered more blood tests and a bone scan to see if there was metastasis. Dr. Bhoopalam concurred with the diagnosis and treatment stating that surgery combined with chemotherapy would be the recommended treatment.

Before the operation Dr. Li did a colonoscopy and Dr. Jack Lea did an endoscopic ultrasound and found that the cancer had gone through several layers of the stomach. They cancelled the major surgery and instead did a minor surgery to insert a port-a-cath, something like a watch pocket in the chest to facilitate a new chemotherapy program, combined with radiation therapy to shrink and kill the cancer. While the minor surgery healed Dr. Reddy planned and did the radiation simulation on the linear accelerator in July 2000.

Radiation and Chemotherapy:

In August 2000 they started the first of 25 radiation treatments or 5 per week for 5 weeks combined with 6 weeks of chemotherapy in 2 three week sessions with IV's of carboplaten, taxol and FU5 every 30 seconds in a fanny pack pump. Al, Patrick and Noreen on 11W displayed a tremendous amount of teamwork to coordinate all this. After the benedryl and zofran wore off in a few days, I became nauseated from the chemotherapy and much sicker than the economy. The stock market slowly got worse and I lost my hair.

There is a lot of cancer research going on and everyday new cures are found. Research gives us hope for the future but it can be expensive. Fortunately George W. Bush promised to spend 91 billion dollars over 10 years in medical research, especially cancer research, in August 2000 before he was elected president.

That summer forest fires destroyed millions of acres of valuable forest- land in the western United States creating lots of pollution in the process. If only the EPA would allow the logging companies to cut trees for logging roads it would help to stop the spread of the large forest fires. This would reduce the cost of lumber and reduce the pollution. In addition the underbrush could be used to create ethanol, an alternate source of energy for the EPA required 15 different grades of summer reformulated gasoline. Thus lowering gas prices instead of fueling forest fires.

Surgery:

After another endoscopic ultrasound, the surgery to remove the stomach was rescheduled for the end of October and beginning of November 2000. Dr. Gerard Aranha was the attending surgeon. Everyone stopped by to visit on the Sunday afternoon and evening before the surgery while I drank a gallon of super-pooper before a liquid supper. The surgery went perfect on Monday. I woke up to say hello to my sister who waited through 6 hours of surgery just to see me in intensive care with all the IV's and heart monitors etc. The next morning the nurses got me up and I was sitting in a chair when Dr. Aranha came in with 7 to 10 doctors. Dr. Aranha explained that the stomach and gall bladder along with several nodes were removed and sent to pathology. I thanked Dr. Aranha and the surgical team of doctors for the great job they did. The doctors explained that I could have something to eat and drink in about a week after everything healed.

The following Tuesday I watched the election returns come in on television. It was very close in Florida. This was tough on the economy. I was released from the hospital on Thursday, November 9, 2000. On Saturday, November 11th, I attended the Veterans Day ceremony at our Park Ridge VFW Post. Our Hospital chairman Frank Toporek had visited me when I was in intensive care and asked what I was doing there. Several weeks later I actually sat down to a real Thanksgiving dinner. General Motors announced the closing of their entire Oldsmobile division and the loss of 37,000 jobs. By the end of the year 450,000 manufacturing jobs would be lost in the US.

Chemotherapy:

In December 2000 and January 2001 I had post surgery chemotherapy treatments with carboplaten and taxol in 2 four-week sessions. Dr. Bhoopalam planned it so that my hair would not fall out till after Christmas. It was shocking watching the TV news with long time companies like Montgomery Wards going out of business, and Motorola laying off 22,000 employees. Perhaps President George W. Bush could help restore consumer confidence by meeting with the leaders of the companies that were planning to close or to lay off employees.

Remission:

Several weeks ago I was told that I was in remission. Isn't that funny, my van failed the EPA Emission Tests 3 times. It has over 100,000 miles on the odometer. They were planning to suspend my license if it did not pass. Before the 4th test I got a letter from the VA Hospital- sort of a doctor's excuse- stating that I needed my van to get to the VA Hospital. Just before the test I also got a new catalytic converter installed and I passed the emission test with a fast pass. The doctor said your cancer is in remission, you passed both the CT Scan and the endoscopic EGD test. Thank you, I said. It's been a long year. It's wonderful to be in remission and be considered a cancer survivor. I hope the economy recovers too.

Reconnaissance and ENERGY VISUALIZATION.
by Stephen E. Weber

APPENDIXES.

Appendix B. Educational Proposals.

COMPUTER SYSTEMS CONSULTING SERVICE –

EDUCATIONAL PROPOSALS:
Celebrating 39th Anniversary. Mr. Weber is a retired CPS teacher and worked for IBM, UIC and Wright College.

1. Proposal for a University Library Computer Network for Northwestern University NOTIS in July 1982.
2. Proposal for a Network Communication Controller using parallel processing for Zenith Electronics in 1982.
3. Combining TV Education with Computer Assisted Instruction by Interfacing a Laser Disk to the IBM PC 1982.
4. Remotely Piloted Miniature Reconnaissance Vehicle (DRONE) Design for NASA at Cal. Tech./ JPL in 1983.
5. On-line Technique Using a Database in an IBM PC to Query and Display Microfiche Data to IBM in 1984.
6. An AT&T Marketing Plan for an IBM Compatible PC Design with DOS running under UNIX in April 1984.
7. Proposal to President Reagan entitled "Let's Put America Back to Work" adjustable rate tariffs January 1985.
8. Proposal submitted to UIC AISS for a University of Illinois Computing **NET**work (UICNET). Sept.1985.
9. A Computer Simulation of UIC LCS Library Computer System using IBM's GPSS, Lotus 123 to UIC in 1986.
10. Enhancing Foreign Policy by Exporting Education Instead of Arms via University Satellite Campuses in 1986.
11. SDI Early Warning System Satellite Computer Network for First Phase Enemy Warhead Tracking June 1987.

12. Computer Vision Vehicle Autopilot System for computer image partitioning and processing to GM 1987.
13. Thesis: "Multiple Asynchronous Parallel Processing System (MAPPS) for Real Time Applications". Apr 1987.
14. A Proposal to Coordinate Research for the Strategic Defense Initiative to Governor Thompson in July 1987.
15. Proposal to Northwestern University: "New Ideas for the Research Parks Success" in March 1988.
16. Proposal to Northern Illinois University for an NIU / Dekalb County Research Park. March 1989.
17. SBIR Proposal to NASA for Satellite Weather Forecasting, Simulation and Control. June 27, 1989.
18. Proposal to President Bush for a Strong Defense and Peacetime Mission for the Military Service. October 1989.
19. Proposals for a "New Concept Vehicle - VSTOL Flying Van" submitted to Ford and General Motors. 1989.
20. Proposal for (CATS) Computer Aided Test and Simulation computer program to Ford, Northrop and GM 1989
21. Relational database University Data Model for Student Registration and Class Scheduling to UIC and NU 2/90.
22. Veterans Membership Information System Database for mail merge implemented with dBASE and Access.
23. Proposal for a Computer Model of the Immune System to Northwestern University Medical School. 3/90.
24. Proposal for a Computer Model of the Immune System to the University of Illinois Medical School. 3/90.
25. Proposal to AMOCO for a Solar Energy Collector - Replacement Panel Study for AMOCO in 4/90.
26. Proposal to SSC Lab. & President Bush for the Superconducting Supercollider cost reduction May 1990.
27. IMS to DB2 translation program. University Data Model: Student Registration/ Class Scheduling May 1990.
28. Proposal to Sears to market internationally by using the Sears Catalog as a market analysis tool July 1991.
29. Proposal to Zenith for a Windows style Operating System with Internet capability for Zenith's HDTV Apr. 95.

30. Letter to President Clinton and the FCC to expedite FCC approval of Zenith's HDTV May 1995.
31. Proposal for University Research Parks to help local business compete with world class competition May 1996.
32. Five proposals to the UIC Information Decision Sciences Dept May 1997 to help get research grants.
33. Medical Expert System Shell Project: Analysis of Blood Chemistry Profiles for UIC EECS 585 April 1998.
34. Proposal for a TOP VIEW Project to provide a fault tolerant GPS display of aircraft location April 1998.
35. IDS 507 Class Project to design, code and test JAVA software agents for database query May 1998.
36. Medical Insurance System using Excel & Access for HMO/PPO proposal preparation and presentation 9/98.
37. Five proposals for aerospace systems to help defense contractors transition to a peacetime economy 11/98.
38. Proposal to expedite solutions for the year 2000 (Y2K) problem to President Clinton in January 1999.
39. Proposal "Analyzing Power Distribution Problems with the Telephone Network" to Commonwealth Edison 1999.
40. Four proposals to help the Wright College Computer Support Dept. in 1996, 1997, 1998 and in May 1999.
41. Proposal "VA Hospital Anti-Aging Program for WWII Veterans" to the VA Hospitals and VFW in October 1999.
42. Proposal to Hillary Clinton and Jimmy Carter for "VA Hospital Anti-Aging Program" for VA budget increase 11/99.
43. Designed tri-fold brochure and wrote over 100 letters to save the VA Teaching Hospitals from closing 2002.
44. Letters to President Bush for "Peace and Economic Prosperity". Let the UN to do their job. January 2003.
45. Wrote and typed the required HIPAA manual for Physicians office and the HIPAA privacy notices. April 2003.

Reconnaissance and ENERGY VISUALIZATION.

By Stephen Weber

APPENDIXES.

Appendix C. **Agent Orange Speech**.

Agent Orange Town Hall Meeting, September 23, 2017, Irish American Heritage Center
Text of Speech

Introduction:
My name is Stephen Weber; I was a SGT in the Air Force from 1963 to 1967. I was an Electronics Tech on the RF-4C Phantom II Reconnaissance Aircraft. I was trained in electronics at Lowry and in RF-4C systems at Shaw. I was sent to Mountain Home to set up the IR/ECM shop, I was sent to Udorn in 1966, the closest Airbase to Hanoi, to do the same for the 432nd Recon Wing. It was the International Headquarters of Air America. They flew unmarked propeller aircraft like the C-123 Providers that sprayed Agent Orange. It was like Terry and the Pirates with the Jolly Green Giants and Pony Express.

Duty Assignment:
My duty was night shift supervisor of the IR shop. We did pre flights, post flights and system maintenance. IR missions were scheduled every hour all night long to monitor the Ho Chi Minh Trail etc. The IR system was like a high definition TV thermal map that showed details at night under camouflage. We had you covered at night. From quantum mechanics, every object emits thermal energy. The video signal was picked up in the IR Receiver with a 4 sided front surface optic scanner rotating at 12,000 RPM. The video in the 12 to 14 micron range was reflected up through 2 sets of folding optics onto a

parabolic reflector and into a mercury doped geranium detector in a cryogenerator cooled to 4 degrees above absolute zero. There are 48,000 line scans per minute. The video signal is amplified with a pre amp and AGC post amp. Then it is processed like a TV signal with horizontal and vertical sweep circuits. The video is recorded on 5 inch film moving at a velocity over altitude rate and time stamped with the longitude and latitude information from the inertial navigation system. The film from each sortie is processed and flown to Ton Son Nut in Saigon for photo intelligence. We had to go around the perimeter of the air base every morning, evening and for midnight chow. The base had metal clad runways that were flooded during the monsoons. The C-123 providers were parked across the tarmac from our IR/ECM/Radar shop. We did not know about Agent Orange dangers. We thought they were spraying for mosquitoes.

Agent Orange Cancer, Surgery, Radiation, Chemotherapy and Claims:
In 1997 while working at Wright College, I was diagnosed with hypothyroidism and prescribed synthroid for the thyroid. In 2000 I collapsed in a running event. This was unusual because I was a long distance runner and had run in 24 marathons. They said that I was anemic. After labs and endoscopic tests, a biopsy indicated a stomach ulcer that was diagnosed as gastric esophageal adeno carcinoma. It was not one of the diseases related to Agent Orange at that time. I was told that I would have to have my entire stomach removed in surgery. They did an endoscopic ultrasound and found that the cancer had gone through several layers of the stomach.

The new approach would be chemotherapy and radiation prior to surgery. In July 2000 they put a port-a-cath in my chest to facilitate chemotherapy. In August I had 25 radiation treatments on the linear accelerator and 3 different chemotherapies. I was bald, nauseated and sicker than a dog. In October 2000 I had surgery to remove my stomach, gall bladder, GE junction and a node on the pancreas. After surgery I had more chemotherapy in January 2001. In June 2001 I was asked to speak at the cancer survivors annual celebration of life ceremony. They filed a claim for me at the Edward Hines Jr. Hospital. It was denied in 2002.

I started substitute teaching in the Chicago Public Schools. After 10 years of teaching, I retired in 2010 at the age of 68. In 2011 I was diagnosed with advanced Prostate Cancer, which is one of the diseases related to Agent Orange. They did a nuclear medicine bone scan and found a lesion on the hip and the pelvis. The cancer had gone outside the prostate. It was Stage 4. I filed a claim. The chemotherapy doctor said the standard of care was androgen deprivation therapy. I wanted radiation seeds, surgery or radiation. I applied for clinical trials at Loyola, Northwestern and the University of Chicago. One of my friends recommended his Urologist outside the VA, who prescribed 45 radiation treatments on the linear accelerator with Calypso and 3 seed implants for perfect alignment. Radiation took the entire summer from June through August of 2011.

In 2012 the VA denied my claim. I filed an appeal. In 2014 I wrote a letter to Senator Dick Durbin and later to Vice President Joe Biden about my claim. Nothing happened. We bought Dr Joel Furman's course on the Immunity Solution and I started juicing with green drinks for bioflavonoids and phytonutrients and polyphenols to fight cancer on the blender and food processor. Agent Orange may have damaged our DNA and genetic makeup as evidenced by the increase in birth defects for children of Vietnam veterans. Finally in 2015 I wrote a letter to President Obama about my claim. It was sent back to the Chicago VA office.

In 2016 I submitted my appeal to the Veterans Legal Support Center and Clinic at the John Marshall Law School. They wrote a legal brief which was submitted to the regional office.

Reconnaissance and ENERGY VISUALIZATION

Stephen Weber

BIBLIOGRAPHY

1. Shrader, Robert. <u>Electronic Communication</u>. New York. McGraw-Hill 1959.

2. Moyer, Richard and Baptiste, Prentice. <u>Science</u>. New York, McGraw-Hill 2000.

3. Myers, David. <u>Psychology</u>, Eighth Edition. New York. Worth Publishers 2007.

4. Cutnell, John and Johnson, Kenneth. <u>Physics</u>, Fourth Edition. New York. John Wiley and Sons. 1998.

5. Hearult, Jeanny. <u>Vision: Images, Signals, and Neural Networks</u>. Grenoble. World Scientific. 2010.

6. Karnow, Stanley. <u>Vietnam, A History</u>. New York. The Viking Press. 1983.

7. Edwards, Bruce H. <u>Prove It: The Art of Mathematical Argument</u>. Chantilly. The Great Courses. 2012.

8. Beard, David and Beard, George. <u>Quantum Mechanics with Applications</u>. Boston. Allyn and Bacon. 1970.

9. Kemper, Dave and Nathan, Ruth. <u>Writers Express</u>. Wilmington, Ma. Write Source. 2000.

10. McGill, Roger. <u>Before, During and After Vietnam. A Cavalryman's Story</u>. Chicago. Self Publishing.

11. "Radiation, Rays." <u>Encyclopedia Britannica.</u> 1957 ed.

12. "Global Warming." <u>Britannica Online Encyclopedia</u>. 2018.

13. "In Memoriam, A Senseless Tragedy." <u>Saturday Evening Post</u>. December 14, 1963, 19-35.

14. Kerod, Robin. <u>The Illustrated History of NASA</u>. New York. Gallery Books. W. H. Smith Publishers. 1986.

15. Walpole, Ronald, Meyers, Raymond. <u>Probability and Statistics for Engineers and Scientists</u>. Third Edition Macmillian Publishing Company. 1985.

16. Attenborough, David. Climate Change- The Facts, PBS Television Special. 2020.

17. Irving, J and Mullineux, N. <u>Mathematics in Physics and Engineering</u>. New York. Academic Press Inc.

18. Callen, Herbert B. <u>Thermodynamics</u>. New York. John Wiley & Sons. Sixth Printing. December 1966.

19. Scientific American, <u>Dark Energy.</u> Spring 2024